Remote Sensing in GEOMORPHOLOGY

Remote Sensing in GEOMORPHOLOGY

Partick Simon

Oxford Book Company
Jaipur ▪ New Delhi

ISBN : 9789380179391

Published in India, 2025

Oxford Book Company
Regd. Off. "Bony Residency", Gate No. 2
Opp. Tilak Public School, Vishweswariya Nagar,
Gopalpura Road, Jaipur - 302018, India
Ph: 0141 - 2761280, Telefax: 2761381
email - oxfordbookjpr@gmail.com
website - oxfordbookcompany.com

Delhi Office: 102, 1st Floor, Satyam House,
4327/3, Ansari Road, Darya Ganj,
New Delhi - 110002, India
Ph: 011-45652440

Distributed By
KEYA INTERNATIONAL PRESS LLP
E-mail: keyainternationalpressllp@gmail.com

Printed and bound in India by Thomson Press Ltd., New Delhi

Contents

Preface .... *vii*

1. Introduction 1
2. Data Acquisition Techniques 56
3. Geographic Information System 94
4. Earth Observation Satellite 133
5. Global Positioning System 172
6. Analytic Signal of Radar Processing 203
7. Fundamentals of 3D Modelling and Visualisation 225

Index 287

Preface

Geomorphology is the study of landforms and landform evolution. Geomorphologists are primarily concerned with the study of earth's surficial features, including their origin and evolution and impact on human activity. Geomorphologists work within disciplines such as physical geography, geology, geodesy, engineering geology, archaeology, climatology, and geotechnical engineering. This broad base of interests contributes to many research styles and interests within the field. In general the structures are much older than the geomorphic forms developed upon them. Such major structural features as folds and faults may go back to far distant periods of diastrophism. Remote sensing and Geomorphology are two disciplines of science, where geomorphology is the science of study of landforms of the earth. It is concerned with pattern of landform, materials and their related processes. Geomorphology as a science deals with evolutionary processes of relief forms of the earth and lead us to understand various processes of evolution of landforms. Remote sensing is the science of acquiring information about an object or phenomena close to the earth's surface by measuring electromagnetic radiation. The rocks of the earth's crust vary in their lithology and structure and hence offer varying degrees of resistance to the gradational processes. Differences in rock composition and structures are reflected not only in regional geomorphic variability but in the local topography as well. The local intensity of particular processes may change notably in response to differences in such factors as temperature, moisture, altitude, exposure, topographic configuration, and the amount and type of vegetal cover. The microclimatic conditions may vary markedly between a valley floor and a hilltop, between a northern and a southern exposure, and between bare ground and that with a heavy vegetal cover. The rate of all weathering, all mass-wasting, all erosion, and all deposition varies appreciably within rather narrow limits in relation to the influence of local conditioning factors.

Remote sensing facilitates change detection by comparing images captured at different time intervals. This technique aids in identifying erosion, deposition, vegetation growth, and other transformative events occurring in geomorphic environments. The utility of remote sensing for geomorphological studies is immense. It allows for the rapid assessment of large areas and for the monitoring of changes to these areas – things that would be impossible to do using field studies alone. A Geographic Information System (GIS) is a computer system that analyzes and displays geographically referenced information. It uses data that is attached to a unique location. Most of the information we have about our world contains a location reference: Where are USGS streamgages located? Where was a rock sample collected? Exactly where are all of a city's fire hydrants? If, for example, a rare plant is observed in three different places, GIS analysis might show that the plants are all on north-facing slopes that are above an elevation of 1,000 feet and that get more than ten inches of rain per year. GIS maps can then display all locations in the area that have similar conditions, so researchers know where to look for more of the rare plants. By knowing the geographic location of farms using a specific fertilizer, GIS analysis of farm locations, stream locations, elevations, and rainfall will show which streams are likely to carry that fertilizer downstream. These are just a few examples of the many uses of GIS in earth sciences, biology, resource management, and many other fields.

This book covers the fundamentals of remote sensing and GIS for engineering students and working engineers.

— *Partick Simon*

Chapter 1

Introduction

Remote sensing is the small- or large-scale acquisition of information of an object or phenomenon, by the use of either recording or real-time sensing devices) that are wireless, or not in physical or intimate contact with the object (such as by way of aircraft, spacecraft, satellite, buoy, or ship). In practice, remote sensing is the stand-off collection through the use of a variety of devices for gathering information on a given object or area. Thus, Earth observation or weather satellite collection platforms, ocean and atmospheric observing weather buoy platforms, the monitoring of a parolee via an ultrasound identification system, Magnetic Resonance Imaging (MRI), Positron Emission Tomography (PET), X-radiation (X-RAY) and space probes are all examples of remote sensing. In modern usage, the term generally refers to the use of imaging sensor technologies including: instruments found in aircraft and spacecraft as well as those used in electrophysiology, and is distinct from other imaging-related fields such as medical imaging.

Overview

There are two main types of remote sensing: passive remote sensing and active remote sensing. Passive sensors detect natural radiation that is emitted or reflected by the object or surrounding area being observed. Reflected sunlight is the most common source of radiation measured by passive sensors. Examples of passive remote sensors include film photography, infrared, charge-coupled devices, and radiometers. Active collection, on the other hand, emits energy in order to scan objects and areas whereupon a sensor then detects and measures the radiation that is reflected or backscattered from the target. RADAR is an example of active remote sensing where the time delay between emission and return is measured, establishing the location, height, speed and direction of an object.

Remote sensing makes it possible to collect data on dangerous or inaccessible areas. Remote sensing applications include monitoring deforestation in areas such as the Amazon Basin, glacial features in Arctic and Antarctic regions, and depth sounding of coastal and ocean depths. Military collection during the cold war made use of stand-off collection of data about dangerous border areas. Remote sensing also replaces costly and slow data collection on the ground, ensuring in the process that areas or objects are not disturbed.

Orbital platforms collect and transmit data from different parts of the electromagnetic spectrum, which in conjunction with larger scale aerial or ground-based sensing and analysis, provides researchers with enough information to monitor trends such as El Nino and other natural long and short term phenomena. Other uses include different areas of the earth sciences such as natural resource management, agricultural fields such as land usage and conservation, and national security and overhead, ground-based and stand-off collection on border areas.

Data Acquisition Techniques

The basis for multispectral collection and analysis is that of examined areas or objects that reflect or emit radiation that stand out from surrounding areas.

Applications of Remote Sensing Data

Radar

Israeli military radar is typical of the type of radar used for air traffic control. The antenna rotates at a steady rate, sweeping the local airspace with a narrow vertical fan-shaped beam, to detect aircraft at all altitudes.

Radar is an object-detection system which uses electromagnetic waves — specifically radio waves — to determine the range, altitude, direction, or speed of both moving and fixed objects such as aircraft, ships, spacecraft, guided missiles, motor vehicles, weather formations, and terrain. The radar dish, or antenna, transmits pulses of radio waves or microwaves which bounce off any object in their path. The object returns a tiny part of the wave's energy to a dish or antenna which is usually located at the same site as the transmitter. Practical radar was developed in secrecy during World War 2 by Britain and other nations. The term *RADAR* was coined in 1940 by the U.S. Navy as an acronym for *radio detection and ranging*. The term *radar* has since entered the English and other languages as the common noun, *radar*, losing all of the capitalization. In the United Kingdom, this technology was initially called RDF (*range and direction finding*), using the same acronym as the one for radio direction finding to conceal its ranging capability.

The modern uses of radar are highly diverse, including air traffic control, radar astronomy, air-defence systems, antimissile systems; nautical radars to locate landmarks and other ships; aircraft anticollision systems; ocean-surveillance systems, outer-space surveillance and rendezvous systems; meteorological precipitation monitoring; altimetry and flight-control systems; guided-missile target-locating systems; and ground-penetrating radar geological observations. Other systems similar to radar have been used in other parts of the electromagnetic spectrum. One example is "lidar", which uses visible light from lasers, rather than radio waves.

History of Radar

The history of radar starts with experiments by Heinrich Hertz in the late 19th century that showed that radio waves were reflected by metallic objects. This possibility was suggested in James Clerk Maxwell's seminal work on electromagnetism. However, it was not until the early 20th century that systems able to use these principles were becoming widely available, and it was German engineer Christian Huelsmeyer who first used them to build a simple ship detection device intended to help avoid collisions in fog (Reichspatent Nr. 165546). Numerous similar systems were developed over the next two decades.

The name radar comes from the acronym RADAR, coined in 1940 by the U.S. Navy for public reference to their highly classified work in Radio Detection And Ranging.

Thus, a true radar system must both detect and provide range (distance) information for a target. Before 1934, no single system gave this performance; some systems were omni-directional and provided ranging information, while others provided rough directional information but not range. A key development was the use of pulses that were timed to provide ranging, which were sent from large antennas that provided accurate directional information. Combining the two allowed for accurate plotting of targets.

Worldwide Differential in Antecedence

In the 1934–1939 period, eight nations developed, independently and in great secrecy, systems of this type: the United States, Great Britain, Germany, the USSR, Japan, the Netherlands, France, and Italy. In addition, Great Britain had shared their basic information with four Commonwealth countries: Australia, Canada, New Zealand, and South Africa, and these countries had also developed indigenous radar systems. During the war, Hungary was added to this list.

Progress during the war was rapid and of great importance, probably one of the decisive factors for the victory of the Allies. By the end of hostilities, the United States, Great Britain, Germany, the USSR, and Japan had a wide diversity of land- and sea-based radars as well as small airborne systems. After the war, radar use was widened to numerous fields including: civil aviation, marine navigation, radar guns for police, meteorology and even medicine.

Significance

The place of radar in the larger story of science and technology is argued differently by different authors. Radar, far more than the atomic bomb, contributed to Allied victory in World War II. Robert Buderi states that it was also the precursor of much modern technology. From a review of his book: ... radar has been the root of a wide range of achievements since the war, producing a veri family tree of modern technologies. Because of radar, astronomers can map the contours of far-off planets, meteorologists can measure rain falling in distant places, air travel is hundreds of times safer than travel by road, long-distance telephone calls are cheaper than postage, computers have become ubiquitous and ordinary people can cook their daily dinners in the time between sitcoms, with what used to be called a *radar range*.

But others think that radar is not so important, since the basic principles were not new. For example, Maurice Pointe, one of the developers in France, said:

Le principe fondamental du radar appartient au patrimoine commun des physiciens : ce qui demeure en fin de compte au credit reel des techniciens se mesure a la realisation effective de materiels operationnels., or roughly

The fundamental principle of the radar belongs to the common patrimony of the physicists : after all, what is left to the real credit of the technicians is measured by the effective realisation of operational materials.

Early Contributors

Heinrich Hertz

In 1887 the German physicist Heinrich Hertz (1857–1894) began experimenting with electromagnetic waves in his laboratory. He found that these waves could be transmitted through different types of materials, and were reflected by others, such as conductors and dielectrics. The existence of electromagnetic waves was predicted earlier by the Scottish physicist James Clerk Maxwell (1831–79), but it was Hertz who first succeeded in generating and detecting what were soon called radio waves.

Guglielmo Marconi

The development of the wireless or radio is often attributed to Guglielmo Marconi (1874–1937). Although he was not the first to "invent" this technology, it might be said that he was the greatest early promoter of practical radio systems and their applications. In a paper read before the Institution of Electrical Engineers in London on March 3, 1899, Marconi described radio beacon experiments he had conducted in Salisbury Plain. Concerning this lecture, in a 1922 paper he wrote:

I also described tests carried out in transmitting a beam of reflected waves across country... and pointed out the possibility of the utility of such a system if applied to lighthouses and lightships, so as to enable vessels in foggy weather to locate dangerous points around the coasts...

It [now] seems to me that it should be possible to design [an] apparatus by means of which a ship could radiate or project a divergent beam of these rays in any desired direction, which rays, if coming across a metallic object, such as another steamer or ship, would be reflected back to a receiver screened from the local transmitter on the sending ship, and thereby immediately reveal the presence and bearing of the other ship in fog or thick weather.

This paper and a speech presenting the paper to a joint meeting of the Institute of Radio Engineers and the American Institute of

Electrical Engineers in New York City on June 20, 1922, is often cited as the seminal event which began widespread interest in the development of radar.

Christian Huelsmeyer

In 1904 Christian Huelsmeyer (1881–1957) gave public demonstrations in Germany and the Netherlands of the use of radio echoes to detect ships so that collisions could be avoided. His device consisted of a simple spark gap used to generate a signal that was aimed using a dipole antenna with a cylindrical parabolic reflector. When a signal reflected from a ship was picked up by a similar antenna attached to the separate coherer receiver, a bell sounded.

During bad weather or fog, the device would be periodically "spun" to check for nearby ships. The apparatus detected presence of ships up to 3 km, and Huelsmeyer planned to extend its capability to 10 km. It did not provide range (distance) information, only warning of a nearby object. He patented the device, called the telemobiloscope, but due to lack of interest by the naval authorities the invention was not put into production.

Huelsmeyer also received a patent amendment for estimating the range to the ship. Using a vertical scan of the horizon with the *telemobiloscope* mounted on a tower, the operator would find the angle at which the return was the most intense and deduce, by simple triangulation, the approximate distance. This is in contrast to the later development of pulsed radar, which determines distance directly.

Nikola Tesla

One of the hundreds of concepts generated by Nikola Tesla (1856–1943) included principles regarding frequency and power levels for primitive radio-location units. In an interview published in *The Electrical Experimenter*, August 1917, Tesla gave the following:

> For instance, by their [standing electromagnetic waves] use we may produce at will, from a sending station, an electrical effect in any particular region of the globe; [with which] we may determine the relative position or course of a moving object, such as a vessel at sea, the distance traversed by the same, or its speed.

Tesla also proposed the use of these standing electromagnetic waves along with pulsed reflected surface waves to determine the relative position, speed, and course of a moving object and other modern concepts of radar. Tesla had first proposed that radio location techniques might help find submarines (for which it is not well-suited) with a fluorescent screen indicator.

United States

In the United States, both the Navy and Army needed means of remotely locating enemy ships and aircraft. In 1930, both services initiated the development of radio equipment that could meet this need. There was little coordination of these efforts; thus, they will be described separately.

United States Navy

In the autumn of 1922, Albert H. Taylor and Leo C. Young at the U.S. Naval Aircraft Radio Laboratory were conducting communication experiments when they noticed that a wooden ship in the Potomac River was interfering with their signals; in effect, they had demonstrated the first multistatic radar, a system that uses separated transmitting and receiving antennas and detects targets due to changes in the signal. In 1930, Lawrence A. Hyland working with Taylor and Young, now at the U.S. Naval Research Laboratory (NRL) in Washington, D.C., used a similar arrangement of radio equipment to detect a passing aircraft. This led to a proposal by Taylor for using this technique for detecting ships and aircraft.

A simple wave-interference apparatus can detect the presence of an object, but it cannot determine its location or velocity. That had to await the invention of pulsed radar, and later, additional encoding techniques to extract this information from a CW signal. When Taylor's group at the NRL were unsuccessful in getting interference radio accepted as a detection means, Young suggested trying pulsing techniques. This would also allow the direct determination of range to the target. The British and the American research groups were independently aware of the advantages of such an approach, but the problem was to develop the timing equipment to make it feasible.

Robert Morris Page was assigned by Taylor to implement Young's suggestion. Page designed a transmitter operating at 60 MHz and pulsed 10 μs in duration and 90 μs between pulses. In December 1934, the apparatus was used to detected a plane at a distance of one mile (1.6 km) flying up and down the Potomac. Although the detection range was small and the indications on the oscilloscope monitor were almost indistinct, it demonstrated the basic concept of a pulsed radar system. Based on this, Page, Taylor, and Young are usually credited with building and demonstrating the world's first true radar.

An important subsequent development by Page was the duplexer, a device that allowed the transmitter and receiver to use the same antenna without over-whelming or destroying the sensitive receiver circuitry. This also solved the problem associated with synchronization of separate transmitter and receiver antennas which is critical to accurate position determination of long-range targets.

The experiments with pulsed radar were continued, primarily in improving the receiver for handling the short pulses. In June 1936, the NRL's first prototype radar system, now operating at 28.6 MHz, was demonstrated to government officials, successfully tracking an aircraft at distances up to 25 miles (40 km). Their radar was based on low frequency signals, at least by today's standards, and thus required large antennas, making it impractical for ship or aircraft mounting. Antenna size is inversely proportional to the operating frequency; therefore, the operating frequency of the system was increased to 200 MHz, allowing much smaller antennas. The frequency of 200 MHz was the highest possible with existing transmitter tubes and other components. The new system was successfully tested at the NRL in April 1937, That same month, the first sea-borne testing was conducted. The equipment was temporarily installed on the USS *Leary*, with a Yagi antenna mounted on a gun barrel for sweeping the field of view.

Based on success of the sea trials, the NRL further improved the system. Page developed the ring oscillator, allowing multiple output tubes and increasing the pulse-power to 15 kW in 5-μs pulses. A 20-by-23-foot (7.0 m), stacked-dipole "bedspring" antenna was used. In laboratory test during 1938, the system, now designated XAF, detected planes at ranges up to 100 miles (160 km). It was installed on the battleship USS *New York* for sea trials starting in January 1939, and became the first operational radio detection and ranging set in the U.S. fleet.

In May 1939, a contract was awarded to RCA for production. Designated CXAM, deliveries started in May 1940. The acronym RADAR was coined from "Radio Detection And Ranging" as a cover reference to this highly secret technology. One of the first CXAM systems was placed aboard the USS *California*, a battleship that was sunk in the Japanese attack on Pearl Harbor on December 7, 1941.

United States Army

As the Great Depression started, economic conditions led the U.S. Army Signal Corps to consolidate its widespread laboratory operations to Fort Monmouth, New Jersey. On June 30, 1930, these were designated the Signal Corps Laboratories (SCL) and Lt. Colonel (Dr.) William R. Blair was appointed the SCL Director.

Among other activities, the SCL was made responsible for research in the detection of aircraft by acoustical and infrared radiation means. Blair had performed his doctoral research in the interaction of electromagnet waves with solid materials, and naturally gave attention to this type of detection. Initially, attempts were made to detect infrared radiation, either from the heat of aircraft engines or as reflected from large searchlights with infrared filters, as well as from radio signals generated by the engine ignition.

Some success was made in the infrared detection, but little was accomplished using radio. In 1932, progress at the Naval Research Laboratory (NRL) on radio interference for aircraft detection was passed on to the Army. While it does not appear that any of this information was used by Blair, the SCL did undertake a systematic survey of what was then known throughout the world about the methods of generating, modulating, and detecting radio signals in the microwave region.

The SCL's first definitive efforts in radio-based target detection started in 1934 when the Chief of the Army Signal Corps, after seeing a microwave demonstration by RCA, suggested that radio-echo techniques be investigated. The SCL called this technique radio position-finding (RPF). Based on the previous investigations, the SCL first tried microwaves. During 1934 and 1935, tests of microwave RPF equipment resulted in Doppler-shifted signals being obtained, initially at only a few hundred feet distance and later greater than a mile. These tests involved a bi-static arrangement, with the transmitter at one end of the signal path and the receiver at the other, and the reflecting target passing through or near the path.

Blair was evidently not aware of the success of a pulsed system at the NRL in December 1934. In an internal 1935 note, Blair had commented:

Consideration is now being given to the scheme of projecting an interrupted sequence of trains of oscillations against the target and attempting to detect the echoes during the interstices between the projections.

In 1936, W. Delmar Hershberger, SCL's Chief Engineer at that time, started a modest project in pulsed microwave transmission. Lacking success with microwaves, Hershberger visited the NRL (where he had earlier worked) and saw a demonstration of their pulsed set. Back at the SCL, he and Robert H. Noyes built an experimental apparatus using a 75 watt, 110 MHz (2.73 m) transmitter with pulse modulation and a receiver patterned on the one at the NRL. A request for project funding was turned down by the War Department, but $75,000 for support was diverted from a previous appropriation for a communication project.

In October 1936, Paul E. Watson became the SCL Chief Engineer and led the project. A field setup near the coast was made with the transmitter and receiver separated by a mile. On December 14, 1936, the experimental set detected at up to 7 mi (11 km) range aircraft flying in and out of New York City.

Work then began on a prototype system. Ralph I. Cole headed receiver work and William S. Marks lead transmitter improvements. Separate receivers and antennas were used for azimuth and elevation detection. Both receiving and the transmitting antennas used large arrays of dipole wires on wooden frames. The system output was intended to aim a searchlight. The first demonstration of the full set was made on the night of May 26, 1937. A bomber was detected and then illuminated by the searchlight. The observers included the Secretary of War, Henry A. Woodring; he was so impressed that the next day orders were given for the full development of the system. Congress gave an appropriation of $250,000.

The frequency was increased to 200 MHz (1.5 m). The transmitter used 16 tubes in a ring oscillator circuit (developed at the NRL), producing about 75 kW peak power. Major James C. Moore was assigned to head the complex electrical and mechanical design of lobe switching antennas. Engineers from Western Electric and Westinghouse were brought in to assist in the overall development. Designated SCR-268, a prototype was successfully demonstrated in late 1938 at Fort Monroe, Virginia. Production of SCR-268 sets was started by Western Electric in 1939, and it entered service in early 1941. Even before the SCR-268 entered service, it had been greatly improved. In a project led by Major (Dr.) Harold A. Zahl, two new configurations evolved – the SCR-270 (mobile) and the SCR-271 (fixed-site). Operation at 106 MHz (2.83 m) was selected, and a single water-cooled tube provided 8 kW (100 kW pulsed) output power. Westinghouse received a production contract, and started deliveries near the end of 1940.

The Army deployed five of the first SCR-270 sets around the island of Oahu in Hawaii. At 7:02 on the morning of December 7, 1941, one of these radars detected a flight of aircraft at a range of 136 miles (219 km) due north. The observation was passed on to an aircraft warning centre where it was misidentified as a flight of U.S. bombers known to be approaching from the mainland. The alarm went unheeded, and at 7:48, the Japanese aircraft first struck at Pearl Harbor.

Great Britain

In 1915, Robert Watson Watt joined the Meteorological Office as a meteorologist, working at an outstation at Aldershot in Hampshire. Over the next 20 years, he studied atmospheric phenomena and developed the use of radio signals generated by lightning strikes to map out the position of thunderstorms. The difficulty in pinpointing the direction of these fleeting signals led to the use of rotating directional antennas, and in 1923 the use of oscilloscopes in order to display the signals. An operator would periodically rotate the antenna and look for "spikes" on the oscilloscope to find the direction of a storm. The operation eventually moved to the outskirts of Slough in Berkshire, and in 1927 formed the Radio Research Station (RRS), Slough, an entity under the Department of Scientific and Industrial Research (DSIR). Watson Watt was appointed the RSS Superintendent.

As war clouds gathered over Great Britain, the likelihood of air raids and the threat of invasion by air and sea drove a major effort in applying science and technology to defence. In November 1934, the Air Ministry established the Committee for Scientific Survey of Air Defence (CSSAD) with the official function of considering "how far recent advances in scientific and technical knowledge can be used to strengthen the present methods of defence against hostile aircraft." Commonly called the "Tizard Committee" after its Chairman, Sir Henry Tizard, this group had a profound influence on technical developments in Great Britain.

H. E. Wimperis, Director of Scientific Research at the Air Ministry and a member of the Tizard Committee, had read about Nikola Tesla's claim of inventing a 'death ray.' Watson Watt, Superintendent of the RSS, Slough, was now well established as an authority in the field of radio, and in January 1935, Wimperis contacted him asking if radio might be used for such a device. After discussing this with his scientific assistant, Arnold F. 'Skip' Wilkins, Watson Watt wrote back that this was unlikely, but added the following comment: *Attention is being turned to the still difficult, but less unpromising, problem of radio detection and numerical considerations on the method of detection by reflected radio waves will be submitted when required.*

Over the following several weeks, Wilkins considered the radio detection problem. He outlined an approach and backed it with detailed calculations of necessary transmitter power, reflection characteristics of an aircraft, and needed receiver sensitivity. Watson Watt sent this information to the Air Ministry on February 12, 1935, in a secret report titled "The Detection of Aircraft by Radio Methods."

Reflection of radio signals was critical to the proposed technique, and the Air Ministry asked if this could be proven. To test this, Wilkins set up receiving equipment in a field near Upper Stowe, Northamptonshire. On February 26, 1935, a Handley Page Heyford bomber flew along a path between the receiving station and the transmitting towers of a BBC shortwave station in nearby Daventry. The aircraft reflected the 6 MHz (49 m) BBC signal, and this was readily detected by Doppler-beat interference at ranges up to 8 mi (13 km). This convincing test, known as the *Daventry Experiment*, was witnessed by a representative from the Air Ministry, and led to the immediate authorization to build a full demonstration system.

Based on pulsed transmission as used for probing the ionosphere, a preliminary system was designed and built at the RRS by the team. Their existing transmitter had a peak power of about 1 kW, and Wilkins had estimated that 100 kW would be needed. Edward George Bowen was added to the team to design and build such a transmitter. Bowens' transmitter operated at 6 MHz (50 m), had a pulse-repetition rate of 25 Hz, a pulse width of 25 μs, and approached the desired power.

Orfordness, a narrow, 19-mile (31 km) peninsula in Suffolk along the coast of the North Sea, was selected as the test site. Here the equipment would be openly operated in the guise of an ionospheric monitoring station. In mid-May, 1935, the equipment was moved to

Orfordness. Six wooden towers were erected, two for stringing the transmitting antenna, and four for corners of crossed receiving antennas. In June, general testing of the equipment began.

On June 17, the first target was detected—a Supermarine Scapa flying boat at 17 mi (27 km) range. It is historically correct that on June 17, 1935, radio-based detection and ranging was first demonstrated in Great Britain. Watson Watt, Wilkins, and Bowen are generally credited with initiating what would later be called radar in this nation.

In December 1935, the British Treasury appropriated £60,000 for a five-station system called Chain Home (CH), covering approaches to the Thames Estuary. The secretary of the Tizard Committee, Albert Percival Rowe, coined the acronym RDF as a cover for the work, meaning Range and Direction Finding but suggesting the already well-known Radio Direction Finding.

Air Ministry

In March 1936 the work at Orfordness was moved to Bawdsey Manor, nearby on the mainland. Until this time, the work had officially still been under the DSIR, but was now transferred to the Air Ministry. At the new Bawdsey Research Station, the CH equipment was assembled as a prototype. There were equipment problems when the Royal Air Force (RAF) first exercised the prototype station in September 1936. These were cleared by the next April, and the Air Ministry started plans for a larger network of stations.

Initial hardware at CH stations was as follows: The transmitter operated on four pre-selected frequencies between 20 and 55 MHz, adjustable within 15 seconds, and delivered a peak power of 200 kW. The pulse duration was adjustable between 5 to 25 μs, with a repetition rate selectable as either 25 or 50 Hz. For synchronization of all CH transmitters, the pulse generator was locked to the 50 Hz of the British power grid. Four 360-foot (110 m) steel towers supported transmitting antennas, and four 240-foot (73 m) wooden towers supported cross-dipole arrays at three different levels. A goniometer was used to improve the directional accuracy from the multiple receiving antennas.

By the summer of 1937, 20 initial CH stations were in check-out operation. A major RAF exercise was performed before the end of the year, and was such a success that £10,000,000 was appropriated by the Treasury for an eventual full chain of coastal stations. At the start of 1938, the RAF took over control of all CH stations, and the network began regular operations.

In May 1938, Rowe replaced Watson Watt as Superintendent at Bawdsey. In addition to the work on CH and successor systems, there was now major work in airborne RDF equipment. This was led by E. G. Bowen and centred on 200-MHz (1.5 m) sets. The higher frequency allowed smaller antennas, appropriate for aircraft installation.

From the initiation of RDF work at Orfordness, the Air Ministry had kept the British Army and the Royal Navy generally informed; this led to both of these forces having their own RDF developments.

British Army

In 1931, at the at the Woolwich Research Station of the Army's Signals Experimental Establishment (SEE), W. A. S. Butement and P. E. Pollard had examined pulsed 600 MHz (50-cm) signals for detection of ships. Although they prepared a memorandum on this subject and performed preliminary experiments, for undefined reasons the War Office did not give it consideration.

As the Air Ministry's work on RDF progressed, Colonel Peter Worlledge of the Royal Engineer and Signals Board met with Watson Watt and was briefed on the RDF equipment and techniques being developed at Orfordness. His report, "The Proposed Method of Aeroplane Detection and Its Prospects," led the SEE to set up an "Army Cell" at Bawdsey in October 1936. This was under E. Talbot Paris and the staff included Butement and Pollard. The Cell's work emphasize two general types of RDF equipment: gun-laying (GL) systems for assisting anti-aircraft guns and searchlights, and coastal-defence (CD) systems for directing coastal artillery and defence of Army bases overseas.

Pollard led the first project, a gun-laying RDF code-named Mobile Radio Unit (MRU). This truck-mounted system was designed a small version of a CH station. It operated at 23 MHz (13 m) with a power of 300 kW.

A single 105-foot (32 m) tower supported a transmitting antenna, as well as two receiving antennas set orthogonally for estimating the signal bearing. In February 1937, a developmental unit detected an aircraft at 60 m (96 km) range. The Air Ministry also adopted this system as a mobile auxiliary to the CH system.

In early 1938, Butement started the development of a CD system based on Bowen's evolving 200-MHz (1.5-m) airborne sets. The transmitter had a 400 Hz pulse rate, a 2-μs pulse width, and 50 kW power (later increased to 150 kW). Although many of Bowen's transmitter and receiver components were used, the system would not be airborne so there were no limitations on antenna size.

Primary credit for introducing beamed RDF systems in Great Britain must be given to Butement. For the CD, he developed a large dipole array, 10 feet (3.0 m) high and 24 feet (7.3 m) wide, giving much narrower beams and higher gain. This could be rotated at a speed up to 1.5 revolutions per minute. For greater directional accuracy, lobe switching on the receiving antennas was adopted. As a part of this development, he formulated the first – at least in Great Britain – mathematical relationship that would later become well known as the "radar range equation."

By May 1939, the CD RDF could detect aircraft flying as low as 500 feet (150 m) and at a range of 25 mi (40 km). With an antenna 60 feet (18 m) above sea level, it could determine the range of a 2,000-ton ship at 24 mi (38 km) and with an angular accuracy of as little as a quarter of a degree.

Royal Navy

Although the Royal Navy maintained close contact with the Air Ministry work at Bawdsey, they chose to establish their own RDF development at the Experimental Department of His Majesty's Signal School (HMSS) in Portsmouth, Hampshire, on the south coast.

The HMSS started RDF work in September 1935. Initial efforts, under R. F. Yeo, were in wavelengths ranging between 75 MHz (4 m) and 1.2 GHz (25 cm). All of the work was under the utmost secrecy; it could not even be discussed with other scientists and engineers at Portsmouth. A 75 MHz range-only set was eventually developed and designated Type 79X. Basic tests were done using a training ship, but the operation was unsatisfactory.

In August 1937, the RDF development at the HMSS changed, with many of their best researchers brought into the activity. John D. S. Rawlinson was made responsible for improving the Type 79X. To increase the efficiency, he decreased the frequency to 43 MHz (7 m). Designated Type 79Y, it had separate, stationary transmitting and receiving antennas.

Prototypes of the Type 79Y air-warning system were successfully tested at sea in early 1938. The detection range on aircraft was between 30 and 50 mi (48 and 80 km), depending on height. The systems were then placed into service in August on the cruiser HMS *Sheffield* and in October on the battleship HMS *Rodney*. These were the first vessels in the Royal Navy with RDF systems.

Germany

A radio-based device for remotely indicating the presence of ships was built in Germany by Christian Hulsmeyer in 1904. Often referred to as the first radar system, this did not directly measure the range (distance) to the target, and thus did not meet the criteria to be given this name.

Over the following three decades in invention-rich Germany, a number of radio-based detection systems were developed, but none were true radars. This situation changed as World War II loomed closer. Developments in three leading industries are described.

GEMA

In the early 1930s, physicist Rudolf Kuhnhold, Scientific Director at the '*Kriegsmarine* (German Navy) *Nachrichtenmittel-Versuchsanstalt* (NVA—Experimental Institute of Communication Systems) in Kiel, was attempting to improve the acoustical methods of underwater detection of ships. He concluded that the desired accuracy in measuring distance to targets could be attained only by using pulsed electromagnetic waves.

During 1933, Kuhnhold first attempted to test this concept with a transmitting and receiving set that operated in the microwave region at 13.5 cm (2.22 GHz). The transmitter used a Barkhausen-Kurz tube (the first microwave generator) that produced only 0.1 watt. Unsuccessful with this, he asked for assistance from Paul-Gunther Erbsloh and Hans-Karl Freiherr von Willisen, amateur radio operators who were developing a VHF system for communications. They enthusiastically agreed, and in January 1934, formed a company, *Gesellschaft fur Elektroakustische und Mechanische Apparate* ((GEMA)), for the effort. From the start, the firm was always called simply GEMA.

Work on a *Funkmessgerat fur Untersuchung* (radio measuring device for reconnaissance) began in earnest at GEMA. Hans Hollmann and Theodor Schultes, both affiliated with the prestigious Heinrich Hertz Institute in Berlin, were added as consultants. The first apparatus used a split-anode magnetron purchased from Philips in the Netherlands. This provided about 70 W at 50 cm (600 MHz), but suffered from frequency instability. Hollmann built a regenerative receiver and Schultes developed Yagi antennas for transmitting and receiving. In June 1934, large vessels passing through the Kiev Harbor were detected by Doppler-beat interference at a distance of about 2 km (1.2 miles). In October, strong reflections were observed from an aircraft that happened to fly through the beam; this opened consideration of targets other than ships.

Kuhnhold then shifted the GEMA work to a pulse-modulated system. A new 50 cm (600 MHz) Philips magnetron with better frequency stability was used. It was modulated with 2- μs pulses at a PRF of 2000 Hz. The transmitting antenna was an array of 10 pairs of dipoles with a reflecting mesh. The wide-band regenerative receiver used Acorn tubes from RCA, and the receiving antenna had three pairs of dipoles and incorporated lobe switching. A blocking device (a duplexer), shut the receiver input when the transmitter pulsed. A Braun tube (a CRT) was used for displaying the range.

The equipment was first tested at a NVA site at the Lubecker Bay near Pelzerhaken. During May 1935, it detected returns from woods across the bay at a range of 15 km (9.3 mi). It had limited success, however, in detecting a research ship, *Welle*, only a short distance away. The receiver was then rebuilt, becoming a super-regenerative set with two intermediate-frequency stages. With this improved receiver, the system readily tracked vessels at up to 8 km (5 mi) range.

In September 1935, a demonstration was given to the Commander-in-Chief of the *Kriegsmarine*. The system performance was excellent; the range was read off the Braun tube with a tolerance of 50 meters (less than 1 percent variance), and the lobe switching allowed a directional accuracy of 0.1 degree. Historically, this marked the first naval vessel equipped with radar. Although this apparatus was not put into production, GEMA was funded to develop similar systems operating around 50 cm (500 MHz). These became the Seetakt for the *Kriegsmarine* and the Freya for the *Luftwaffe* (German Air Force).

Kuhnhold remained with the NVA but also consulted with GEMA. He is considered by many in Germany as the Father of Radar. During

1933-6, Hollmann wrote the first comprehensive treatise on microwaves, *Physik und Technik der ultrakurzen Wellen* (Physics and Technique of Ultrashort Waves), Springer 1938.

Telefunken

In 1933, when Kuhnhold at the NVA was first experimenting with microwaves, he had sought information from Telefunken on microwave tubes. (Telefunken was the largest supplier of radio products in Germany) There, Wilhelm Tolme Runge had told him that no vacuum tubes were available for these frequencies. In fact, Runge was already experimenting with high-frequency transmitters and had Telefunken's tube department working on cm-wavelength devices.

In the summer of 1935, Runge, now Director of Telefunken's Radio Research Laboratory, initiated an internally funded project in radio-based detection. Using Barkhausen-Kurz tubes, a 50 cm (600 MHz) receiver and 0.5-W transmitter were built. With the antennas placed flat on the ground some distance apart, Runge arranged for an aircraft to fly overhead and found that the receiver gave a strong Doppler-beat interference signal.

Runge, now with Hans Hollmann as a consultant, continued in developing a 1.8 m (170 MHz) system using pulse-modulation. Wilhelm Stepp developed a transmit-receive device (a duplexer) for allowing a common antenna. Stepp also code-named the system *Darmstadt* after his home town, starting the practice in Telefunken of giving the systems names of cities. The system, with only a few watts transmitter power, was first tested in February 1936, detecting an aircraft at about 5 km (3 mi) distance. This led the *Luftwaffe* to fund the development of a 50 cm (600 MHz) gun-laying system, the *Wurzburg*.

Lorenz

Since before the First World War, Standard Elektrik Lorenz had been the main supplier of communication equipment for the German military and was the main rival of Telefunken. In late 1935, when Lorenz found that Runge at Telefunken was doing research in radio-based detection equipment, they started a similar activity under Gottfried Muller. A pulse-modulated set called *Einheit fur Abfragung* (DFA- Device for Detection) was built. It used a type DS-310 tube (similar to the Acorn) operating at 70 cm (430 MHz) and about 1 kW power, it had identical transmitting and receiving antennas made with rows of half-wavelength dipoles backed by a reflecting screen.

In early 1936, initial experiments gave reflections from large buildings at up to about 7 km (4 mi). The power was doubled by using two tubes, and in mid-1936, the equipment was set up on cliffs near Keil, and good detections of ships at 7 km (4 mi) and aircraft at 4 km (2.5 mi) were attained.

The success of this experimental set was reported to the *Kriegsmarine*, but they showed no interest; they were already fully engaged with GEMA for similar equipment. Also, because of extensive agreements between Lorenz and many foreign countries, the naval authorities had reservations concerning the company handling classified work. The DFA was then demonstrated to the *Heer* (German Army), and they contracted with Lorenz for developing *Kurfurst* (Cure prince), a system for supporting *Flugzeugabwehrkanone* (Flak, anti-aircraft guns).

USSR

In 1895, Alexander Stepanovich Popov, a physics instructor at the Imperial Russian Navy school in Kronstadt, developed an apparatus using a coherer tube for detecting distant lighting strikes. The next year, he added a spark-gap transmitter and demonstrated the first radio communication set in Russia. During 1897, while testing this in communicating between two ships in the Baltic Sea, he took note of an interference beat caused by the passage of a third vessel. In his report, Popov wrote that this phenomenon might be used for detecting objects, but he did nothing more with this observation.

In a few years following the 1917 Russian Revolution and the establishment the Union of Soviet Socialist Republics (USSR or Soviet Union) in 1924, Germany's *Luftwaffe* had aircraft capable of penetrating deep into Soviet territory. Thus, the detection of aircraft at night or above clouds was of great interest to the *Voiska Protivo-vozdushnoi aborony* (PVO, Air Defence Forces) of the *Raboche-Krest'yanskaya Krasnaya Armiya* (RKKA, Workers'–Peasants' Red Army).

The PVO depended on optical devices for locating targets, and had physicist Pavel K. Oshchepkov conducting research in possible improvement of these devices. In June 1933, Oshchepkov changed his research from optics to radio techniques and started the development of a *razvedyvlatl'naya elektromagnitnaya stantsiya* (reconnaissance electromagnetic station). In a short time, Oshchepkov was made responsible for a PVO *experino-tekknicheskii sektor* (technical expertise sector) devoted to *radiolokatory* (radio-location) techniques as well as heading a Special Construction Bureau (SCB) in Leningrad (formerly St. Petersberg).

Radio-Location Beginnings

The *Glavnoe artilkeriisko upravlenie* (GAU, Main Artillery Administration) was considered the "brains" of the Red Army. It not only had competent engineers and physicists on its central staff, but also had a number of scientific research institutes. Thus, the GAU was also assigned the aircraft detection problem, and Lt. Gen. M. M. Lobanov was placed in charge.

After examining existing optical and acoustical equipment, Lobanov also turned to radio-location techniques. For this he approached the *Tsentral'naya radiolaboratoriya* (TsRL, Central Radio Laboratory) in Leningrad. Here, Yu. K. Korovin was conducting research on VHF communications, and had built a 50 cm (600 MHz), 0.2 W transmitter using a Barkhausen-Kurz tube. For testing the concept, Korovin arranged the transmitting and receiving antennas along the flight path of an aircraft. On January 3, 1934, a Doppler signal was received by reflections from the aircraft at some 600 m range and 100–150 m altitude.

For further research in detection methods, a major conference on this subject was arranged for the PVO by the *Rossiiskaya Akademiya Nauk* (RAN, Russian Academy of Sciences). The conference was held in Leningrad in mid-January, 1934, and chaired by Abram Fedorovich Ioffe, Director of the Leningrad Physical-Technical Institute (LPTI). Ioffe was generally considered the top Russian physicist of his time. All types of detection techniques were discussed, but radio-location received the greatest attention. To distribute the conference findings to a wider audience, the proceedings were published the following month in a journal. This included all of the then-existing information on radio-location in the USSR, available (in Russian language) to researchers in this field throughout the world.

Recognizing the potential value of radio-location to the military, the GAU made a separate agreement with the Leningrad Electro-Physics Institute (LEPI), for a radio-location system. This technical effort was led by B. K. Shembel. The LEPI had built a transmitter and receiver to study the radio-reflection characteristics of various materials and targets. Shemlbel readily made this into an experimental bi-static radio-location system called *Bistro* (Rapid). The *Bistro* transmitter, operating at 4.7 m (64 MHz), produced near 200 W and was frequency-modulated by a 1 kHz tone. A fixed transmitting antenna gave a broad coverage of what was called a *radiozkzn* (radio screen). A regenerative receiver, located some distance from the transmitter, had a dipole antenna mounted on a hand-driven reciprocating mechanism. An aircraft passing into the screened zone would reflect the radiation, and the receiver would detect the Doppler-interference beat between the transmitted and reflected signals.

Bistro was first tested during the summer of 1934. With the receiver up to 11 km away from the transmitter, the set could only detect an aircraft entering a screen at about 3 km (2 mi) range and a under 1,000 m. With improvements, it was believed to have a potential range of 75 km, and five sets were ordered in October for field trials. *Bistro* is often cited as the USSR's first radar system; however, it was incapable of directly measuring range and thus could not be so classified.

LEPI and TsRL were both made a part of *Nauchno-issledovatel institut-9* (NII-9, Scientific Research Institute #9), a new GAU organization opened in Leningrad in 1935. Mikhail A. Bonch-Bruyevich, a renowned radio physicist previously with TsRL and the University of Leningrad, was named the NII-9 Scientific Director.

Research on magnetrons began at Kharkov University in Ukraine during the mid-1920s. Before the end of the decade this had resulted in publications with worldwide distribution, such as the German journal *Annalen der Physik* (*Annals of Physics*). Based on this work, Ioffe recommended that a portion of the LEPI be transferred to the city of Kharkov, resulting in the Ukrainian Institute of Physics and Technology (LIPT) being formed in 1930. Within the LIPT, the Laboratory of Electromagnetic Oscillations (LEMO), headed by Abram A. Slutskin, continued with magnetron development. Led by Aleksandr S. Usikov, a number of advanced segmented-anode magnetrons evolved. (It is noted that these and other early magnetrons developed in the USSR suffered from frequency instability, a problem in their use in Soviet radar systems.)

In 1936, one of Usikov's magnetrons producing about 7 W at 18 cm (1.7 GHz) was used by Shembel at the NII-9 as a transmitter in a *radioiskatel* (radio-seeker) called *Burya* (Storm). Operating similarly to *Bistro*, the range of detection was about 10 km, and provided azimuth and elevation coordinates estimated to within 4 degrees. No attempts were made to make this intro a pulsed system, thus, it could not provide range and was not qualified to be classified as a radar. It was, however, the first microwave radio-detection system.

While work by Shembel and Bonch-Bruyevich on continuous-wave systems was taking place at NII-9, Oshehepkov at the SCB and V. V. Tsimbalin of Ioffe's LPTI were pursuing a pulsed system. In 1936, they built a radio-location set operating at 4 m (75 MHz) with a peak-power of about 500 W and a 10-μs pulse duration. Before the end of the year, tests using separated transmitting and receiving sites resulted in an aircraft being detected at 7 km. In April 1937, with the peak-pulse power increased to 1 kW and the antenna separation also increased, test showed a detection range of near 17 km at a height of 1.5 km. Although a pulsed system, it was not capable of directly providing range – the technique of using pulses for determining range had not yet been developed.

Pre-War Radio-Location Systems

In June 1937, all of the work in Leningrad on radio-location suddenly stopped. The infamous Great Purge of dictator Joseph Stalin swept over the military high commands and its supporting scientific community. The PVO chief was executed. Oshchepkov, charged with "high crime," was sentenced to 10 years at a Gulag penal labour camp. NII-9 as an organization was saved, but Shenbel was dismissed and Bonch-Bruyevich was named the new director.

The *Nauchnoissledovatel skii ispytalel nyi institut suyazi RKKA* (NIIIS-KA, Scientific Research Institute of Signals of the Red Army), had initially opposed research in radio-location, favouring instead acoustical techniques. However, this portion of the Red Army gained power as a result of the Great Purge, and did an about face, pressing hard for speedy development of radio-location systems. They took over Oshchepkov's laboratory and were made responsible for all existing and future agreements for research and factory production.

Writing later about the Purge and subsequent effects, General Lobanov commented that it led to the development being placed under a single organization, and the rapid reorganization of the work to accomplish needed results.

At Oshchepkov's former laboratory, work with the 4 m (75 MHz) pulsed-transmission system was continued by A. I. Shestako. Through pulsing, the transmitter produced a peak power of 1 kW, the highest level thus far generated. In July 1938, a fixed-position, bi-static experimental system detected an aircraft at about 30 km range at heights of 500 m, and at 95 km range, for high-flying targets at 7.5 km altitude. The system was still incapable of directly determining the range. The project was then taken up by Ioffe's LPTI, resulting in the development of a mobile system designated *Redut* (Redoubt). An arrangement of new transmitter tubes was used, giving near 50 kW peak-power with a 10 μs pulse-duration. Yagi antennas were adopted for both transmitting and receiving.

The *Redut* was first field tested in October 1939, at a site near Sevastopol, a port in Ukraine on the coast of the Black Sea. This testing was in part to show the NKKF (Soviet Navy) the value of early-warning radio-location for protecting strategic ports. With the equipment on a cliff about 160 meters above sea level, a flying boat was detected at ranges up to 150 km. The Yagi antennas were spaced about 1,000 meters; thus, close coordination was required to aim them in synchronization.

At the NII-9 under Bonch-Bruyevich, scientists developed two types of very advanced microwave generators. In 1938, a linear-beam, velocity-modulated vacuum tube (a klystron) was developed by N. D. Devyatkov, based on designs from Kharkpv. This device produced about 25 W at 15–18 cm (2.0–1.7 GHz) and was later used in experimental systems. Devyatkov followed this with a simpler, single-resonator device (a reflex klystron). At this same time, D. E. Malyarov and N. F. Alekseyev were building a series of magnetrons, also based on designs from Kharkov; the best of these produced 300 W at 9 cm (3 GHz).

Also at NII-9, D. S. Stogov was placed in charge of the improvements to the *Bistro* system. Redesignated as *Reven* (Rhubarb), it was tested in August 1938, but was only marginally better than the predecessor. With additional minor operational improvements, it was made into a mobile system called *Radio Ulavlivatel Samoletov* (RUS, Radio Catcher of Aircraft), soon designated as *RUS-1*. This continuous-wave, bi-static system had a truck-mounted transmitter operating at 4.7 m (64 MHz) and two truck-mounted receivers.

Although the *RUS-1* transmitter was in a cabin on the rear of a truck, the antenna had to be strung between external poles anchored to the ground. A second truck carrying the electrical generator and other equipment was backed against the transmitter truck. Two receivers were used, each in a truck-mounted cabin with a dipole antenna on a rotatable pole extended overhead. In use, the receiver trucks were placed about 40 km apart; thus, with two positions, it would be possible to make a rough estimate of the range by triangulation on a map.

The *RUS-1* system was tested and put into production in 1939, then entered service in 1940, becoming the first deployed radio-location system in the Red Army. A total of about 45 *RUS-1* systems were built at the Svetlana Factory in Leningrad before the end of 1941, and deployed along the western USSR borders and in the Far East. Without direct ranging capability, however, the military found the *RUS-1* to be of little value.

Even before the demise of efforts in Leningrad, the NIIIS-KA had contracted with the UIPT in Kharkov to investigate a pulsed radio-location system for anti-aircraft applications. This led the LEMO, in March 1937, to start an internally funded project with the code name *Zenit* (a popular football team at the time). The transmitter development was led by Usikov, supplier of the magnetron used earlier in the *Burya*. For the *Zenit*, Usikov used a 60 cm (500 MHz) magnetron pulsed at 10–20 μs duration and providing 3 kW pulsed power, later increased to near 10 kW. Semion Braude led the development of a superheterodyne receiver using a tunable magnetron as the local oscillator. The system had separate transmitting and receiving antennas set about 65 m apart, built with dipoles backed by 3-meter parabolic reflectors.

Zenit was first tested in October 1938. In this, a medium-sized bomber was detected at a range of 3 km. The testing was observed by the NIIIS-KA and found to be sufficient for starting a contracted effort. An agreement was made in May 1939, specifying the required performance and calling for the system to be ready for production by 1941. The transmitter was increased in power, the antennas had selsens added to allow them to track, and the receiver sensitivity was improved by using an RCA 955 acorn triode as the local oscillator.

A demonstration of the improved *Zenit* was given in September 1940. In this, it was shown that the range, altitude, and azimuth of an aircraft flying at heights between 4,000 and 7,000 meters could be determined at up to 25 km distance. The time required for these measurements, however, was about 38 seconds, far too long for use by anti-aircraft batteries. Also, with the antennas aimed at a low angle, there was a dead zone of some distance caused by interference from ground-level reflections. While this performance was not satisfactory for immediate gun-laying applications, it was the first full three-coordinate radio-location system in the Soviet Union and showed the way for future systems.

Work at the LEMO continued on *Zenit*, particularly in converting it into a single-antenna system designated *Rubin*. This effort, however, was disrupted by the invasion of the USSR by Germany in June 1941. In a short while, the development activities at Kharkov were ordered to be evacuated into the Far East. The research efforts in Leningrad were similarly dispersed.

After eight years of effort by highly qualified physicists and engineers, the USSR entered World War II without a fully developed and fielded radar system.

Japan

As a sea-faring nation, Japan had an early interest in wireless (radio) communications. The first known use of wireless telegraphy in warfare at sea was by the Imperial Japanese Navy, in defeating the Russian Imperial Fleet in 1904. There was an early interest in equipment for radio direction-finding, for use in both navigation and military surveillance. The Imperial Navy developed an excellent receiver for this purpose in 1921, and soon most of the Japanese warships had this equipment.

In the two decades between the two World Wars, radio technology in Japan made advancements on a par with that in the western nations. There were often impediments, however, in transferring these advancements into the military. For a long time, the Japanese had believed they had the best fighting capability of any military force in the world. The military leaders, who were then also in control of the government, sincerely felt that the weapons, aircraft, and ships that they had built were fully sufficient and, with these as they were, the Japanese Army and Navy were invincible. In 1936, Japan joined Nazi Germany and Fascist Italy in a Tripartite Pact.

Technology Background

Radio engineering was strong in Japan's higher education institutions, especially the Imperial (government-financed) universities.

This included undergraduate and graduate study, as well as academic research in this field. Special relationships were established with foreign universities and institutes, particularly in Germany, with Japanese teachers and researchers often going overseas for advanced study.

The academic research tended toward the improvement of basic technologies, rather than their specific applications. There was considerable research in high-frequency and high-power oscillators, such as the magnetron, but the application of these devices was generally left to industrial and military researchers.

One of Japan's best-known radio researchers in the 1920s-30s era was Professor Hidetsugu Yagi. After graduate study in Germany, England, and America, Yagi joined Tohoku University where his research centred on antennas and oscillators for high-frequency communications. A summary of the radio research work at Tohoku University was contained in a 1928 seminal paper by Yagi.

Jointly with Shintaro Uda, one Yagi's first doctoral students, a radically new antenna emerged. It had a number of parasitic elements (directors and reflectors) and would come to be known as the Yagi-Uda or Yagi antenna. A U.S. patent, issued in May 1932, was assigned to RCA. To this day, this is the most widely used directional antenna worldwide.

The cavity magnetron was also of interest to Yagi. This HF (~10-MHz) device had been invented in 1921 by Albert W. Hull at General Electric, and Yagi was convinced that it could function in the VHF or even the UHF region. In 1927, Kinjiro Okabe, another of Yagi's early doctoral students, developed a split-anode device that ultimately generated oscillations at wavelengths down to about 12 cm (2.5 GHz).

Researchers at other Japanese universities and institutions also started projects in magnetron development, leading to improvements in the split-anode device. These included Kiyoshi Morita at the Tokyo Institute of Technology, and Tsuneo Ito at Tokoku University.

Shigeru Nakajima at Japan Radio Company (JRC) saw a commercial potential of these devices and began the further development and subsequent very profitable production of magnetrons for the medical dielectric heating (diathermy) market. The only military interest in magnetrons was shown by Yoji Ito at the Naval Technical Research Institute (NTRI).

The NTRI was formed in 1922, and became fully operational in 1930. Located at Meguro, Tokyo, near the Tokyo Institute of Technology, first-rate scientists, engineers, and technicians were engaged in activities ranging from designing giant submarines to building new radio tubes. Included were all of the precursors of radar, but this did not mean that the heads of the Imperial Navy accepted these accomplishments.

In 1936, Tsuneo Ito (no relationship to Yoji Ito) developed an 8-split-anode magnetron that produced about 10 W at 10 cm (3 GHz). Based on its appearance, it was named *Tachibana* (or Mandarin, an orange citrus fruit). Tsuneo Ito also joined the NTRI and continued his research on magnetrons in association with Yoji Ito. In 1937, they developed the technique of coupling adjacent segments (called push-pull), resulting in frequency stability, an extremely important magnetron breakthrough.

By early 1939, NTRI/JRC had jointly developed a 10-cm (3-GHz), stable-frequency Mandarin-type magnetron (No. M3) that, with water cooling, could produce 500-W power. In the same time period, magnetrons were built with 10 and 12 cavities operating as low as 0.7 cm (40 GHz). The configuration of the M3 magnetron was essentially the same as that used later in the magnetron developed by Boot and Randall at Birmingham University in early 1940, including the improvement of strapped cavities. Unlike the high-power magnetron in Great Britain, however, the initial device from the NTRI generated only a few hundred watts.

In general, there was no lack of scientific and engineering capabilities in Japan; their warships and aircraft clearly showed high levels of technical competency. They were ahead of Great Britain in the development of magnetrons, and their Yagi antenna was the world standard for VHF systems. It was simply that the top military leaders failed to recognize how the application of radio in detection and ranging – what was often called the Radio Range Finder (RRF) – could be of value, particularly in any offensive role; offense not defence, totally dominated their thinking.

Imperial Army

In 1938, engineers from the Research Office of Nippon Electric Company (NEC) were making coverage tests on high-frequency transmitters when rapid fading of the signal was observed. This occurred whenever an aircraft passed over the line between the transmitter and receiving meter. Masatsugu Kobayashi, the Manager of NEC's Tube Department, recognized that this was due to the beat-frequency interference of the direct signal and the Doppler-shifted signal reflected from the aircraft.

Kobayashi suggested to the Army Science Research Institute that this phenomenon might be used as an aircraft warning method. Although the Army had rejected earlier proposals for using radio-detection techniques, this one had appeal because it was based on an easily understandable method and would require little developmental cost and risk to prove its military value. NEC assigned Kinji Satake of their Research Institute to develop a system called the Bi-static Doppler Interference Detector (BDID).

For testing the prototype system, it was set up on an area recently occupied by Japan along the coast of China. The system operated between 4.0-7.5 MHz (75–40 m) and involved a number of widely spaced stations; this formed a radio screen that could detect the presence (but nothing more) of an aircraft at distances up to 500 km (315 mi). The BDID was the Imperial Army's first deployed radio-based detection system, placed into operation in early 1941.

A similar system was developed by Satake for the Japanese homeland. Information centres received oral warnings from the operators at BDID stations, usually spaced between 65 and 240 km (40 and 150 mi). To reduce homing vulnerability – a great fear of the military – the transmitters operated with only a few watts power. Although originally intended to be temporary until better systems were available, they remained in operation throughout the war. It was not until after the start of war the Imperial Army had equipment that could be called radar.

Imperial Navy

In the mid-1930s, some of the technical specialists in the Imperial Navy became interested in the possibility of using radio to detect aircraft. For consultation, they turned to Professor Yagi who, then the Director of the Radio Research Laboratory at Osaka Imperial University. Yagi suggested that this might be done by examining the Doppler frequency-shift in a reflected signal.

Funding was provided to the Osaka Laboratory for experimental investigation of this technique. Kinjiro Okabe, the inventor of the split-anode magnetron and who had followed Yagi to Osaka, led the effort. Theoretical analyses indicated that the reflections would be greater if the wavelength was approximately the same as the size of aircraft structures. Thus, a VHF transmitter and receiver with Yagi antennas separated some distance were used for the experiment.

In 1936, Okabe successfully detected a passing aircraft by the Doppler-interference method; this was the first recorded demonstration in Japan of aircraft detection by radio. With this success, Okabe's research interest switched from magnetrons to VHF equipment for target detection. This, however, did not lead to any significant funding. The top levels of the Imperial Navy believed that any advantage of using radio for this purpose were greatly outweighed by enemy intercept and disclosure of the sender's presence.

Historically, warships in formation used lights and horns to avoid collision at night or when in fog. Newer techniques of VHF radio communications and direction-finding might also be used, but all of these methods were highly vulnerable to enemy interception. At the NTRI, Yoji Ito proposed that the UHF signal from a magnetron might be used to generate a very narrow beam that would have a greatly reduced chance of enemy detection.

Development of microwave system for collision avoidance started in 1939, when funding was provided by the Imperial Navy to JRC for preliminary experiments. In a cooperative effort involving Yoji Ito of the NTRI and Shigeru Nakajima of JRC, an apparatus using a 3-cm (10-GHz) magnetron with frequency modulation was designed and built. The equipment was used in an attempt to detect reflections from tall structures a few kilometers away. This experiment gave poor results, attributed to the very low power from the magnetron.

The initial magnetron was replaced by one operating at 16 cm (1.9 GHz) and with considerably higher power. The results were then much better, and in October 1940, the equipment obtained clear echoes from a ship in Tokyo Bay at a distance of about 10 km (6 mi). There was still no commitment by top Japanese naval officials for using this technology aboard warships. Nothing more was done at this time, but late in 1941, the system was adopted for limited use.

In late 1940, Japan arranged for two technical missions to visit Germany and exchange information about their developments in military technology. Commander Yoji Ito represented the Navy's interest in radio applications, and Lieutenant Colonel Kinji Satake did the same for the Army. During a visit of several months, they exchanged significant general information, as well as limited secret materials in some technologies, but little directly concerning radio-detection techniques. Neither side even mentioned magnetrons, but the Germans did apparently disclose their use of pulsed techniques.

After receiving the reports from the technical exchange in Germany, as well as intelligence reports concerning the success of Great Britain with firing using RDF, the Naval General Staff reversed itself and tentatively accepted pulse-transmission technology. On August 2, 1941, even before Yoji Ito returned to Japan, funds were allocated for the initial development of pulse-modulated radars. Commander Chuji Hashimoto of the NTRI was responsible for initiating this activity.

A prototype set operating at 4.2 m (71 MHz) and producing about 5 kW was completed on a crash basis. With the NTRI in the lead, the firm NEC and the Research Laboratory of Japan Broadcasting Corporation (NHK) made major contributions to the effort.

Kenjiro Takayanagi, Chief Engineer of NHK's experimental television station and called "the father of Japanese television," was especially helpful in rapidly developing the pulse-forming and timing circuits, as well as the receiver display. In early September 1941, the prototype set was first tested; it detected a single bomber at 97 km (60 mi) and a flight of aircraft at 145 km (91 mi).

The system, Japan's first full Radio Range Finder (RRF – radar), was designated Mark 1 Model 1. Contracts were given to three firms for serial production; NEC built the transmitters and pulse modulators, Japan Victor the receivers and associated displays, and Fuji Electrical the antennas and their servo drives. The system operated at 3.0 m (100 MHz) with a peak-power of 40 kW. Dipole arrays with matte+-type reflectors were used in separate antennas for transmitting and receiving.

In November 1941, the first manufactured RRF was placed into service as a land-based early-warning system at Katsuura, Chiba, a town on the Pacific coast about 100 km (62 mi) from Tokyo. A large system, it weighed close to 8,700 kg (19,000 lb). The detection range was about 130 km (81 mi) for single aircraft and 250 km (156 mi) for groups.

Netherlands

Early radio-based detection in the Netherlands were along two independent lines: one a microwaves system at the firm Philips and the other a VHF system at a laboratory of the Armed Forces.

The Philips Company in Eindhoven, Netherlands, operated *Natuurkundig Laboratorium* (NatLab) for fundamental research related to its products. NatLab researcher Klass Posthumus developed a magnetron split into four elements. In developing a communication system using this magnetron, C.H.J.A. Stall was testing the transmission by using parabolic transmitting and receiving antennas set side-by-side, both aimed at a large plate some distance away. To overcome frequency instability of the magnetron, pulse modulation was used. It was found that the plate reflected a strong signal.

Recognizing the potential importance of this as a detection device, NatLab arranged a demonstration for the *Koninklijke Marine* (Royal Netherlands Navy). This was conducted in 1937 across the entrance to the main naval port at Marsdiep. Reflections from sea waves obscured the return from the target ship, but the Navy was sufficiently impressed to initiate sponsorship of the research. In 1939, an improved set was demonstrated at Wijk aan Zee, detecting a vessel at a distance of 3.2 km (2.0 mi).

A prototype system was built by Philips, and plans were started by the firm Nederlandse Seintoestellen Fabriek (a Philips subsidiary) for building a chain of warning stations to protect the primary ports. Some field testing of the prototype was conducted, but the project was discontinued when Germany invaded the Netherlands on May 10, 1940. Within the NatLab, however, the work was continued in great secrecy until 1942.

During the early 1930s, there were widespread rumours of a "death ray" being developed. The Dutch Parliament set up a Committee for the Applications of Physics in Weaponry under G.J. Elias to examine this potential, but the Committee quickly discounted death rays. The Committee did, however, establish the *Laboratorium voor Fysieke Ontwikkeling* (LFO, Laboratory for Physical Development), dedicated to supporting the Netherlands Armed Forces.

Operating in great secrecy, the LFO opened a facility called the *Meetgebouw* (Measurements Building) located on the Plain of Waalsdorp. In 1934, J.L.W.C. von Weiler joined the LFO and, with

S.G. Gratama, began research on a 1.25-m (240-MHz) communication system to be used in artillery spotting.

In 1937, while tests were being conducted on this system, a passing flock of birds disturbed the signal. Realizing that this might be a potential method for detecting aircraft, the Minister of War ordered continuation of the experiments. Weiler and Gratama set about developing a system for directing searchlights and aiming anti-aircraft guns.

The experimental "electrical listening device" operated at 70 cm (430 MHz) and used pulsed transmission at an RPF of 10 kHz. A transmit-receive blocking circuit was developed to allow a common antenna. The received signal was displayed on a CR tube with a circular time base. This set was demonstrated to the Army in April 1938 and detected an aircraft at a range of 18 km (11 mi). The set was rejected, however, because it could not withstand the harsh environment of Army combat conditions.

The Navy was more receptive. Funding was provided for final development, and Max Staal was added to the team. To maintain secrecy, they divided the development into parts. The transmitter was built at the Delft Technical College and the receiver at the University of Leiden. Ten sets would be assembled under the personal supervision of J.J.A. Schagen van Leeuwen, head of the firm Hazemeijer Fabriek van Signaalapparaten.

The prototype had a peak-power of 1 kW, and used a pulse length of 2 to 3 μs with a 10- to 20 kHz PRF. The receiver was a super-heterodyne type using Acorn tubes and a 6 MHz IF stage. The antenna consisted of 4 rows of 16 half-wave dipoles backed by a 3- by 3-meter mesh screen. The operator used a bicycle-type drive to rotate the antenna, and the elevation could be changed using a hand crank.

Several sets were completed, and one was put into operation on the Malievelt in The Hague just before the Netherlands fell to Germany in May 1940.

The set worked well, spotting enemy aircraft during the first days of fighting. To prevent capture, operating units and plans for the system were destroyed. Von Weiler and Max Staal fled to England aboard one of the last ships able to leave, carrying two disassembled sets with them. Later, Gratama and van Leeuwen also escaped to England.

France

In 1927, French physicists Camille Gutton and Emile Pierret experimented with magnetrons and other devices generating wavelengths going down to 16 cm. Camille's son, Henri Gutton, was with the *Compagnie Generale de Telegraphie Sans Fil* (CSF) where he and Robert Warneck improved his father's magnetrons. Following reports made by the U.S. Naval Research Laboratory concerning detection by interference methods, in 1934 Henri Gutton started the development of a radio-detection apparatus using the short-wavelength tubes.

Emile Girardeau, the head of the CSF, recalled in testimony that they were at the time intending to build radio-detection systems "*conceived according to the principles stated by Tesla.*" In 1934, CSF submitted a patent application for a device for detecting obstacles using continuous radiation of ultra-short wavelengths produced by a magnetron. Since this device could not directly measure distance, it was not what would later be called radar, but it was the first patent of an operational radio-detection apparatus using centimetric wavelengths.

The system was tested in late 1934 aboard the cargo ship *Oregon*, with two transmitters working at 80 cm and 16 cm wavelengths. Coastlines were detected from a range of 10-12 nautical miles. The shortest wavelength was chosen for the final design, which equipped the liner *SS Normandie* as early as mid-1935 for operational use. These were still CW systems and depended on Doppler interference for detection.

In late 1937, Maurice Elie at SFR developed a means of pulse-modulating transmitter tubes. This led to a new 16-cm system with a peak power near 10 W and a pulse width of 6 μs. The Navy set up tests in early 1939 and detected large vessels at 10 km, but it was unable to detect aircraft and was thus not accepted by the military. French and U.S. patents were filed in December 1939. The system was planned to be sea-tested aboard the *Normandie*, but this was cancelled at the outbreak of war.

Pierre David at the *Laboratori National de Radioelectricite* (National Laboratory of Radioelectricity, LNR) experimented with reflected radio signals at about a meter wavelength. Starting in 1931, he observed that aircraft caused interference to the signals. The LNR then initiated research on a detection technique called *barrage electromagnetique* (electromagnetic curtain). While this could indicate the general location of penetration, precise determination of direction and speed was not possible.

In 1936, the *Defense Aerienne de Territoire* (Defence of Air Territory), ran tests on David's electromagnetic curtain. In the tests, the system detected most of the entering aircraft, but too many were missed. As the war grew closer, the need for an aircraft detection was critical. David realized the advantages of a pulsed system, and in October 1938 he designed a 50 MHz, pulse-modulated system with a peak-pulse power of 12 kW. This was built by the firm SADIR.

France declared war on Germany on September 1, 1939, and there was a great need for an early-warning detection system. The SADIR system was taken to near Tulon, and detected and measured the range of invading aircraft as far as 55 km (34 mi).

The SFR pulsed system was set up near Paris where it detected aircraft at ranges up to 130 km (81 mi). To further improve the SFR system, Henri Gitton and Warneck developed a 500-cm (600-MHz) magnetron with an oxide-coated cathode that produced 500 W. In May 1940, just before the Germans arrived, several of the new magnetrons were taken to Great Britain where their oxide-coated cathods were incorporated to improve the Boots and Randall magnetron.

Italy

Guglielmo Marconi initiated the research in Italy on radio-based detection technology. In 1933, while participating with his Italian firm in experiments with a 600 MHz communications link across Rome, he noted transmission disturbances caused by moving objects adjacent to its path.

This led to the development at his laboratory at Cornegliano of a 330-MHz (0.91-m) CW Doppler detection system that he called *radioecometro*. Barkhausen-Kurz tubes were used in both the transmitter and receiver.

In May 1935, Marconi demonstrated his system to the Fascist dictator Benito Mussolini and members of the military General Staff; however the output power was insufficient for military use. While Marconi's demonstration raised considerable interest, little more was done with his apparatus. Mussolini directed that radio-based detection technology be further developed, and it was assigned to the *Regio*

Instituto Electrotecnico e delle Comunicazioni (RIEC, Royal Institute for Electro-technics and Communications). The RIEC had been established in 1916 on the campus of the Italian Naval Academy in Livorno. Lieutenant Ugo Tiberio, a physics and radio-technology instructor at the Academy, was assigned to head the project on a part-time basis.

Tiberio prepared a report on developing an experimental apparatus that he called *telemetro radiofonico del rivelatore* (RDT, Radio-Detector Telemetry). The report, submitted in mid-1936, included what was later known as the radar range equation. When the work got underway, Nello Carrara, a civilian physics instructor who had been doing research at the RIEC in microwaves, was added to be responsible for developing the RDT transmitter.

Before the end of 1936, Tiberio and Carrara had demonstrated the EC-1, the first Italian RDT system. This had an FM transmitter operating at 200 MHz (1.5 m) with a single parabolic cylinder antenna. It detected by mixing the transmitted and the Doppler-shifted reflected signals, resulting in an audible tone.

The EC-1 did not provide a range measurement; to add this capability, development of a pulsed system was initiated in 1937. Captain Alfeo Brandimarte joined the group and primarily designed the first pulsed system, the EC-2. This operated at 175 MHz (1.7 m) and used a single antenna made with a number of equi-phased dipoles. The detected signal was intended to be displayed on an oscilloscope. There were many problems, and the system never reached the testing stage. Work then turned to developing higher power and operating frequencies. Carrara, in cooperation with the firm FIVRE, developed a magnetron-like device. This was composed of a pair of triodes connected to a resonate cavity and produced 10 kW at 425 MHz (70 cm). It was used in designing two versions of the EC-3, one for shipboard and the other for coastal defence.

Italy, joining Germany, entered WWII in September 1939 without an operational RDT. A breadboard of the EC-3 was built and tested from atop a building at the Academy, but most RDT work was stopped as direct support of the war took priority.

Other Nations

In early 1939, the British Government invited representatives from the most technically advanced Commonwealth Nations to visit

England for briefings and demonstrations on the highly secret RDF (radar) technology. Based on this, RDF developments were started in Australia, Canada, New Zealand, and South Africa by September 1939. In addition, this technology was independently developed in Hungary early in the war period.

In Australia, the Radiophysics Laboratory was established at Sydney University under the Council for Scientific and Industrial Research; John H. Piddington was responsible for RDF development. The first project was a 200-MHz (1.5-m) shore-defence system for the Australian Army. Designated ShD, this was first tested in September 1941, and eventually installed at 17 ports. Following the Japanese attack on Pearl Harbor, the Royal Australian Air Force urgently needed an air-warning system, and Piddington's team, using the ShD as a basis, put the AW Mark I together in five days. It was being installed in Darwin, Northern Territory, when Australia received the first Japanese attack on February 19, 1942. A short time later, it was converted to a light-weight transportable version, the LW-AW Mark II; this was used by the Australian forces, as well as the U.S. Army, in early island landings in the South Pacific.

The early RDF developments in Canada were at the Radio Section of the National Research Council of Canada. Using commercial components and with essentially no further assistance from Great Britain, John Tasker Henderson led a team in developing the Night Watchman, a surface-warning system for the Royal Canadian Navy to protect the entrance to the Halifax Harbour. Successfully tested in July 1940, this set operated at 200 MHz (1.5 m), had a 1 kW output with a pulse length of 0.5 μs, and used a relatively small, fixed antenna. This was followed by a ship-borne set designated Surface Warning 1st Canadian (SW1C) with the antenna hand-rotated; this was first tested at sea in mid-May 1941, For coastal defence by the Canadian Army, a 200 MHz set with a transmitter similar to the Night Watchman was developed. Designated CD, it used a large, rotating antenna atop a 70-foot (21 m) wooden tower. The CD was put into operation in January 1942.

Ernest Marsden represented New Zealand at the briefings in England, and then established two facilities for RDF development – one in Wellington at the Radio Section of the Central NZ Post Office, and another at Canterbury University College in Christchurch. Charles N. Watson-Munro led the development of land-based and airborne sets at Wellington, while Frederick W. G. White led the development of shipboard sets at Christchurch.

Before the end of 1939, the Wellington group had converted an existing 180-MHz (1.6-m), 1 kW transmitter to produce 2-μs pulses and tested it to detect large vessels at up to 30 km; this was designated CW (Coastal Watching). A similar set, designated CD (Coast Defence) used a CRT for display and had lobe-switching on the receiving antenna; this was deployed in Wellington in late 1940. A partially completed ASV 200 MHz set was brought from Great Britain by Marsden, and another group at Wellington built this into an aircraft set for the Royal New Zealand Air Force; this was first flown in early 1940. At Christchurch, there was a smaller staff and work went slower, but by July 1940, a 430-MHz (70-cm), 5 kW set was tested. Two types, designated SW (Ship Warning) and SWG (Ship Warning, Gunnery), were placed into service by the Royal New Zealand Navy starting in August 1941.

South Africa did not have a representative at the 1939 meetings in England, but in mid-September, as Ernest Marsden was returning by ship to New Zealand, Basil F. J. Schonland came aboard and received three days of briefings. Schonland, a world authority on lightning and Director of the Bernard Price Institute of Geophysics at Witwatersrand University, immediately started an RDF development using amateur radio components and Institute's lightning-monitoring equipment. Designated JB (for Johannesburg), the 90-MHz (3.3-m), 500-W mobile system was tested in November 1939, just two months after its start. The prototype was operated in Durban before the end of 1939, detecting ships and aircraft at distances up to 80 km, and by the next March a system was fielded by anti-aircraft brigades of the South African Defence Force.

In Hungary, Zoltan Lajos Bay was a Professor of Physics at the Technical University of Budapest as well as the Research Director of Egyesult Izzolampa (IZZO), a radio and electrical manufacturing firm. In late 1942, IZZO was directed by the Minister of Defence to develop a radio-location system. Using journal papers on ionospheric measurements for information on pulsed transmission, Bay developed a system called *Sas* (Eagle) around existing communications hardware.

The *Sas* operated at 120 MHz (2.5 m) and was in a cabin with separate transmitting and receiving dipole arrays attached; the assembly was all on a rotatable platform. According to published records, the system was tested in 1944 atop Mount Janos and had a range of "better than 500 km." A second *Sas* was installed at another location. There is no indication that either *Sas* installation was ever in regular service. After the war, Bay used a modified *Sas* to successfully bounce a signal off the moon.

World War II Radar

At the start of World War II in September 1939, both the United Kingdom and Germany knew of each other's ongoing efforts in radio navigation and its countermeasures – the "Battle of the beams". Also, both nations were generally aware of, and intensely interested in, the other's developments in radio-based detection and tracking, and engaged in an active campaign of espionage and false leaks about their respective equipment.

By the time of the Battle of Britain, both sides were deploying range and direction-finding units (radars) and control stations as part of integrated air defence capability. However, the German *Funkmessgerät* (radio measuring device) systems could not assist in an offensive role and was thus not supported by Adolf Hitler. Also, the *Luftwaffe* did not sufficiently appreciate the importance of British Range and Direction Finding (RDF) stations as part of RAF's air defence capability, contributing to their failure.

While the United Kingdom and Germany led in pre-war advances in the use of radio for detection and tracking of aircraft, there were also developments in the United States, the Soviet Union, and Japan. Wartime systems in all of these nations will be summarized. The acronym RADAR (for Radio Detection And Ranging) was coined by the U.S. Navy in 1940, and the subsequent name "radar" was soon widely used.

Post-WWII Radar

World War II, which gave impetus to the great surge in radar development, ended between the Allies and Germany in May 1945, followed by Japan in August. With this, radar activities in Germany and Japan ceased for a number of years. In other countries, particularly the United States, Great Britain, and the USSR, the politically unstable post-war years saw continued radar improvements for military applications. In fact, these three nations all made significant efforts in bringing scientists and engineers from Germany to work in their weapon programs; in the U.S., this was under Operation Paperclip.

Even before the end of the war, various project directed toward non-military applications of radar and closely related technologies were initiated. The US Army Air Forces and the British RAF had made wartime advances in using radar for handling aircraft landing, and this was rapidly expanded into the civil sector. The field of radio astronomy was one of the related technologies; although discovered before the war, it immediately flourished in the late 1940s with many scientists around the world establishing new careers based on their radar experience. Four techniques, highly important in post-war radars, were matured in the late 1940s-early 1950s: pulse Doppler, monopulse, phased array, and synthetic aperture; the first three were known and even used during wartime developments, but were matured later.

- Pulse-Doppler radar (often known as moving target indication or MTI), uses the Doppler-shifted signals from targets to better detect moving targets in the presence of clutter.
- Monopulse radar (also called simultaneous lobing) was conceived by Robert Page at the NRL in 1943. With this, the system derives error-angle information from a single pulse, greatly improving the tracking accuracy.
- Phased-array radar has the many segments of a large antenna separately controlled, allowing the beam to be quickly directed. This greatly reduces the time necessary to change the beam direction from one point to another, allowing almost simultaneous tracking of multiple targets while maintaining overall surveillance.
- Synthetic-aperture radar (SAR), was invented in the early 1950s at Goodyear Aircraft Corporation. Using a single, relatively small antenna carried on an aircraft, a SAR combines the returns from each pulse to produce a high-resolution image of the terrain comparable that obtained by a much larger antenna. SAR has wide applications, particularly in mapping and remote sensing.

One of the early applications of digital computers was in switching the signal phase in elements of large phased-array antennas. As smaller computers came into being, these were quickly applied to digital signal processing using algorithms for improve radar performance. Other advances in radar systems and applications in the decades following WWII are far too many to be included herein. The following sections are intended to provide representative samples.

Military Radars

In the United States, the Rad Lab at MIT officially closed at the end of 1945. The Naval Research Laboratory (NRL) and the Army's Evans Signal Laboratory continued with new activities in centimetre radar development. The United States Air Force (USAF) – separated from the Army in 1946 – concentrated radar research at their Cambridge Research Centre (CRC) at Hanscom Field, Massachusetts. In 1951, MIT opened the Lincoln Laboratory for joint developments with the CRC. While the Bell Telephone Laboratories embarked on major communications upgrades, they continued with the Army in radar for their ongoing Nike air-defence program.

In Great Britain, the RAF's Telecommunications Research Establishment (TRE) and the Army's Radar Research and Development Establishment (RRDE) both continued at reduced levels at Malvern, Worcestershire, then in 1953 were combined to form the Radar Research Establishment. In 1948, all of the Royal Navy's radio and radar R&D activities were combined to form the Admiralty Signal and Radar Establishment, located near Portsmouth, Hampshire. The USSR, although devastated by the war, immediately embarked on the development of new weapons, including radars.

During the Cold War period following WWII, the primary "axis" of combat shifted to lie between the United States and the Soviet Union. By 1949, both sides had nuclear weapons carried by bombers. To provide early warning of an attack, both deployed huge radar networks of increasing sophistication at evermore remote locations. In the West, the first such system was the Pinetree Line, deployed across Canada in the early 1950s, backed up with radar pickets on ships and oil platforms off the east and west coasts.

The Pinetree Line initially used vintage pulsed radars and was soon supplemented with the Mid-Canada Line (MCL). Soviet technology improvements made these Lines inadequate and, in a construction project involving 25,000 persons, the Distant Early Warning Line (DEW Line) was completed in 1957. Stretching from Alaska to Baffin Island and covering over 6,000 miles (9,600 km), the DEW Line consisted of 63 stations with AN/FPS-19 high-power, pulsed, L-Band radars, most augmented by AN/FPS-23 pulse-Doppler systems. The Soviet Unit tested its first Intercontinental Ballistic Missile (ICBM) in August 1957, and in a few years the early-earning role was passed almost entirely to the more capable DEW Line.

Both the U.S. and the Soviet Union then had ICBMs with nuclear warheads, and each began the development of a major anti-ballistic missile (ABM) system. In the USSR, this was the Fakel V-1000, and for this they developed powerful radar systems. This was eventually deployed around Moscow as the A-35 anti-ballistic missile system, supported by radars designated by NATO as the Cat House, Dog House, and Hen House.

In 1957, the U.S. Army initiated an ABM system first called Nike-X; this passed through several names, eventually becoming the Safeguard Program. For this, there was a long-range Perimeter Acquisition Radar (PAR) and a shorter-range, more precise Missile Site Radar (MSR).

The PAR was housed in a 128-foot (39 m)-high nuclear-hardened building with one face sloping 25 degrees facing north. This contained 6,888 antenna elements separated in transmitting and receiving phased arrays. The L-Band transmitter used 128 long-life travelling-wave tubes (TWTs), having a combined power in the megawatt range The PAR could detect incoming missiles outside the atmosphere at distances up to 1,800 miles (2,000 km).

The MSR had an 80-foot (24 m), truncated pyramid structure, with each face holding a phased-array antenna 13 feet (4.0 m) in diameter and containing 5,001 array elements used for both transmitting and receiving. Operating in the S-Band, the transmitter used two klystrons functioning in parallel, each with megawatt-level power. The MSR could search for targets from all directions, acquiring them at up to 300-miles (480-km) range. One Safeguard site, intended to defend Minuteman ICBM missile silos near the Grand Forks AFB in North Dakota, was finally completed in October 1975, but the U.S. Congress withdrew all funding after it was operational but a single day. During the following decades, the U.S. Army and the U.S. Air Force developed a variety of large radar systems, but the long-serving BTL gave up military development work in the 1970s.

A modern radar developed by of the U.S. Navy that should be noted is the AN/SPY-1. First fielded in 1973, this S-Band, 6 MW system has gone through a number of variants and is a major component of the Aegis Combat System. An automatic detect-and-track system, it is computer controlled using four complementary three-dimensional passive electronically scanned array antennas to provide hemispherical coverage.

Radar signals, travelling with line-of-sight propagation, normally have a range to ground targets limited by the visible horizon, or less than about 10 miles (16 km). Airborne targets can be detected by ground-level radars at greater ranges, but, at best, several hundred miles. Since the beginning of radio, it had been known that signals of appropriate frequencies (3 to 30 MHz) could be "bounced" from the ionosphere and received at considerable distances. As long-range bombers and missiles came into being, there was a need to have radars give early warnings at great ranges. In the early 1950s, a team at the Naval Research Laboratory came up with the Over-the-Horizon (OTH) radar for this purpose.

To distinguish targets from other reflectors, it was necessary to use a phase-Doppler system. Very sensitive receivers with low-noise amplifiers had to be developed. Since the signal going to the target and returning had a propagation loss proportional to the range raised to the fourth power, a powerful transmitter and large antennas were required. A digital computer with considerable capability (new at that time) was necessary for analysing the data. In 1950, their first experimental system was able to detect rocket launches 600 miles (960 km) away at Cape Canaveral, and the cloud from a nuclear explosion in Nevada 1,700 miles (2,700 km) distant.

In the early 1970s, a joint American-British project, code named Cobra Mist, used a 10-MW OTH radar at Orfordness (the birthplace of British radar), England, in an attempt to detect aircraft and missile launchings over the Western USSR. Because of US-USSR ABM agreements, this was abandoned within two years. In the same time period, the Soviets were developing a similar system; this successfully detected a missile launch at 2,500 km (1,600 miles). By 1976, this had matured into an operational system named *Duga* ("Arc" in English), but known to western intelligence as Steel Yard and called Woodpecker by radio amateurs and others who suffered from its interference – the transmitter was estimated to have a power of 10 MW. Australia, Canada, and France also developed OTH radar systems.

With the advent of satellites with early-warning capabilities, the military lost most of its interest in OTH radars. However, in recent years, this technology has been reactivated for detecting and tracking ocean shipping in applications such as maritime reconnaissance and drug enforcement. Systems using an alternate technology have also been developed for over-the-horizon detection. Due to diffraction, electromagnetic surface waves are scattered to the rear of objects, and these signals can be detected in a direction opposite from high-powered transmissions. Called OTH-SW (SW for Surface Wave), Russia is using such a system to monitor the Sea of Japan, and Canada has a system for coastal surveillance.

Civil Aviation Radars

The post-war years saw the beginning of a revolutionary development in Air Traffic Control (ATC) – the introduction of radar. In 1946, the Civil Aeronautics Administration (CAA) unveiled an experimental radar-equipped tower for control of civil flights. By 1952, the CAA had begun its first routine use of radar for approach and departure control.

Four years later, it placed a large order for long-range radars for use in *en route* ATC; these had the capability, at higher altitudes, to see aircraft within 200 nautical miles (370 km). In 1960, it became required for aircraft flying in certain areas to carry a radar transponder that identified the aircraft and helped improve radar performance. Since 1966, the responsible agency has been called the Federal Aviation Administration (FAA).

A Terminal Radar Approach Control (TRACON) is an ATC facility usually located within the vicinity of a large airport. In the US Air Force it is known as RAPCON (Radar Approach Control), and in the US Navy as a RATCF (Radar Air Traffic Control Facility). Typically, the TRACON controls aircraft within a 30 to 50 nautical mile (56 to 93 km) radius of the airport at an altitude between 10,000 to 15,000 feet (3,000 to 4,600 m). This uses one or more Airport Surveillance Radars (ASR-7, 8, & 9), sweeping the sky once every few seconds.

The Digital Airport Surveillance Radar (DASR) is a newer TRACOM radar system, replacing the old analog systems with digital technology. The civilian nomenclature for this radar is the ASR-11, and AN/GPN-30 is used by the military. Two radar systems are included. The primary is an X-Band (~2.8 GHz) system with 25 kW pulse power.

It provides 3-D tracking of target aircraft and also measures rainfall intensity. The secondary is a P-Band (~1.05 GHz) system with a peak-power of about 25 kW. It uses a transponder set to interrogate aircraft and receive operational data. The antennas for both systems rotate atop a tall tower.

Weather Radar

The U.S. Weather Bureau was established in 1870 with a specific mission:

To provide for taking meteorological observations at the military stations in the interior of the continent and at other points in the States and Territories...and for giving notice on the northern [Great] Lakes and on the seacoast by magnetic telegraph and marine signals, of the approach and force of storms.

In fulfilling this mission, the Weather Bureau established the first radio network in the United States – possibly in the world – shortly after the Galveston hurricane of 1900, one of the greatest natural disaster in America's history. The continued use of technologies carried into the post-war era with the use of radar for meteorological monitoring. Today, the name "radar" is, to most people, tied to weather reports and warnings via the television. The WSR-1 (Weather Surveillance Radar-1) was one of the first weather radars. This was a modified version of the AN/APS-2F radar, which the Weather Bureau acquired from the Navy. The WSR-1A, WSR-3, and WSR-4 were also variants of this radar. The Weather Bureau later became the National Weather Service (NWS), and is now an agency of the National Oceanic and Atmospheric Administration (NOAA).

The WSR-57 (Weather Surveillance Radar – 1957) was the first weather radar designed specifically for a national warning network. Using WWII technology based on vacuum tubes, it gave only coarse reflectivity data and no velocity information. Operating at 2.89 GHz (S-Band), it had a peak-power of 410 kW and a maximum range of about 580 mi (915 km). AN/FPS-41 was the military designation for the WSR-57.

The next significant change was the WSR-74 series, beginning operations in 1974. There were two types: the WSR-74S, for replacements and filling gaps in the WSR-57 national network, and the WSR-74C, primarily for local use. Both were transistor-based, and their primary technical difference was indicated by the letter, S-Band (better suited for long range) and C-Band, respectively.

Until the 1990s, there were 128 of the WSR-57 and WSR-74 model radars were spread across the country in the NWS radar network. They were gradually replaced by the WSR-88D (Weather Surveillance Radar – 1988, Doppler), commonly known as NEXRAD (Next

Generation Radar). There are now over 150 NEXRAD sites networked in the United States and selected overseas locations.

The WSR-88D has a klystron-based 2.7-3.0-GHz (S-Band) transmitter producing 750 kW peak-power into a parabolic-dish antenna. The PRF is selectable between 318 and 1304 Hz. The system provides reflectivity data at 0.25 km by 0.5 degree to 460-km (290-mi) range, and Doppler data at 0.25 km by 0.5 degree to 230-km (170-mi) range.

In 1998, the NWS began using the Advanced Weather Interactive Processing System (AWIPS). This is a technologically advanced, interactive computer system that integrates data from the NEXRAD network with satellite imagery and surface observations for preparing forecasts and issuing warnings. With upgrades, this provides the images now well known to weather television watchers.

Mapping Radar

The plan position indicator, dating from the early days of radar and still the most common type of display, provides a map of the targets surrounding the radar location. If the radar antenna on an aircraft is aimed downward, a map of the terrain is generated, and the larger the antenna, the greater the image resolution. After centimetre radar came into being, downward-looking radars – the H2S (L-Band) and H2X (C-Band) – provided real-time maps used by the U.S. and Great Britain in bombing runs over Europe at night and through dense clouds.

In 1951, Carl Wiley led a team at Goodyear Aircraft Corporation (later Goodyear Aerospace) in developing a technique for greatly expanding and improving the resolution of radar-generated images. Called synthetic aperture radar (SAR), an ordinary-sized antenna fixed to the side of an aircraft is used with highly complex signal processing to give an image that would otherwise require a much larger, scanning antenna; thus, the name synthetic aperture. As each pulse is emitted, it is radiated over a lateral band onto the terrain. The return is spread in time, due to reflections from features at different distances. Motion of the vehicle along the flight path gives the horizontal increments. The amplitude and phase of returns are combined by the signal processor using Fourier transform techniques in forming the image. The overall technique is closely akind to optical holography. Through the years, many variations of the SAR have been made with diversified applications resulting. In initial systems, the signal processing was too complex for on-board operation; the signals were recorded and processed later. Processors using optical techniques were then tried for generating real-time images, but advances in high-speed electronics now allow on-board processes for most applications. Early systems gave a resolution in tens of meters, but more recent airborne systems provide resolutions to about 10 cm. Current ultra-wideband systems have resolutions of a few millimetres.

Other Radars and Applications

There are many other post-war radar systems and applications. Only a few will be noted.

Radar Gun

The most widespread radar device today is undoubtedly the radar gun. This is a small, usually hand-held, Doppler radar that is used to detect the speed of objects, especially trucks and automobiles in regulating traffic, as well as pitched baseballs, runners, or other moving objects in sports. This device can also be used to measure the surface speed of water and continuously manufactured materials. A radar gun does not return information regarding the object's position; it uses the Doppler effect to measure the speed of a target. First developed in 1954, most radar guns operate with very low power in the X or Ku Bands. Some use infrared radiation or laser light; these are usually called LIDAR. A related technology for velocity measurements in flowing liquids or gasses is called laser Doppler velocimetry; this technology dates from the mid-1960s.

Impulse Radar

As pulsed radars were initially being developed, the use of very narrow pulses was examined. The pulse length governs the accuracy of distance measurement by radar – the shorter the pulse, the greater the precision. Also, for a given pulse repetition frequency (PRF), a shorter pulse results in a higher peak power. Harmonic analysis shows that the narrower the pulse, the wider the band of frequencies that contain the energy, leading to such systems also being called wide-band radars. In the early days, the electronics for generating and receiving these pulses was not available; thus, essentially no applications of this were initially made.

By the 1970s, advances in electronics led to renewed interest in what was often called short-pulse radar. With further advances, it became practical to generate pulses having a width on the same order as the period of the RF carrier (T = 1/f). This is now generally called impulse radar.

The first significant application of this technology was in ground-penetrating radar (GPR). Developed in the 1970s, GPR is now used for structural foundation analysis, archeological mapping, treasure hunting, unexploded ordinance identification, and other shallow investigations. This is possible because impulse radar can concisely locate the boundaries between the general media (the soil) and the desired target. The results, however, are non-unique and are highly dependant upon the skill of the operator and the subsequent interpretation of the data.

In dry or otherwise favourable soil and rock, penetration up to 300 feet (91 m) is often possible. For distance measurements at these short ranges, the transmitted pulse is usually only one radio-frequency cycle in duration; With a 100 MHz carrier and a PRF of 10 kHz (typical parameters), the pulse duration is only 10 ns (nanosecond). leading to the "impulse" designation. A variety of GPR systems are commercially available in back-pack and wheeled-cart versions with pulse-power up to a kilowatt.

With continued development of electronics, systems with pulse durations measured in picoseconds became possible. Applications are as varied as security and motion sensors, building stud-finders, collision-warning devices, and cardiac-dynamics monitors. Some of these devices are match-box sized, including a long-life power source.

Radar Astronomy

As radar was being developed, astronomers considered its application in making observations of the Moon and other near-by extraterrestrial objects. In 1944, Zoltán Lajos Bay had this as a major objective as he developed a radar in Hungary. Sadly, his radar telescope was taken away by the conquering Soviet army and had to be rebuilt, thus delaying the experiment. Under Project Diana conducted by the Army's Evans Signal Laboratory in New Jersey, a modified SCR-271 radar (the fixed-position version of the SCR-270) operating at 110 MHz with 3 kW peak-power, was used in receiving echoes from the Moon on January 10, 1946. Zoltán Bay accomplished this on the following February 6.

Radio astronomy also had its start following WWII, and many scientists involved in radar development then entered this field. A number of radio observatories were constructed during the following years; however, because of the additional cost and complexity of involving transmitters and associated receiving equipment, very few were dedicated to radar astronomy. In fact, essentially all major radar astronomy activities have been conducted as adjuncts to radio astronomy observatories.

The radio telescope at the Arecibo Observatory, opened in 1963, is the largest in the world. Owned by the U.S. National Science Foundation and contractor operated, it is used primarily for radio astronomy, but equipment is available for radar astronomy. This includes transmitters operating at 47 MHz, 439 MHz, and 2.38 GHz, all with very-high pulse power. It has a 305-m (1,000-ft) primary reflector fixed in position; the secondary reflector is on tracks to allow precise pointing to different parts of the sky. Many significant scientific discoveries have been made using the Arecibo radar telescope, including mapping of surface roughness of Mars and observations of Saturns and its largest moon, Titan. In 1989, the observatory radar-imaged an asteroid for the first time in history.

Several spacecraft orbiting the Moon, Mercury, Venus, Mars, and Saturn have carried radars for surface mapping; a ground-penetration radar was carried on the Mars Express mission. Radar systems on a number of aircraft and orbiting spacecraft have mapped the entire Earth for various purposes; on the Shuttle Radar Topography Mission, the entire planet was mapped at a 30-m resolution.

The Jodrell Bank Observatory, an operation of the University of Manchester in Great Britain, was originally started by Bernard Lovell to be a radar astronomy facility. It initially used a war-surplus GL-II radar system operating at 71 MHz (4.2 m). The first observations were of ionized trails in the Geminids meteor shower during December 1945. While the facility soon evolved to become the third largest radio observatory in the world, some radar astronomy continued. The largest (250-ft or 76-m in diameter) of their three fully steerable radio telescopes became operational just in time to radar track *Sputnik 1*, the first artificial satellite, in October 1957.

Applications of Radar

The information provided by radar includes the bearing and range (and therefore position) of the object from the radar scanner. It is thus used in many different fields where the need for such positioning is crucial. The first use of radar was for military purposes: to locate air, ground and sea targets. This evolved in the civilian field into applications for aircraft, ships and roads.

In aviation, aircraft are equipped with radar devices that warn of obstacles in or approaching their path and give accurate altitude readings. They can land in fog at airports equipped with radar-assisted ground-controlled approach (GCA) systems, in which the plane's flight is observed on radar screens while operators radio landing directions to the pilot.

Marine radars are used to measure the bearing and distance of ships to prevent collision with other ships, to navigate and to fix their position at sea when within range of shore or other fixed references such as islands, buoys, and lightships. In port or in harbour, vessel traffic service radar systems are used to monitor and regulate ship movements in busy waters. Police forces use radar guns to monitor vehicle speeds on the roads.

Radar has invaded many other fields. Meteorologists use radar to monitor precipitation. It has become the primary tool for short-term weather forecasting and to watch for severe weather such as thunderstorms, tornadoes, winter storms, precipitation types, etc. Geologists use specialised ground-penetrating radars to map the composition of the Earth's crust.

Principles

A radar system has a transmitter that emits radio waves called *radar signals* in predetermined directions. When these come into contact with an object they are usually reflected and/or scattered in many directions. Radar signals are reflected especially well by materials of considerable electrical conductivity—especially by most metals, by seawater, by wet land, and by wetlands. Some of these make the use of radar altimeters possible. The radar signals that are reflected back towards the transmitter are the desirable ones that make radar work. If the object is *moving* either closer or farther away, there is a slight change in the frequency of the radio waves, due to the Doppler effect.

Radar receivers are usually, but not always, in the same location as the transmitter. Although the reflected radar signals captured by the receiving antenna are usually very weak, these signals can be strengthened by the electronic amplifiers that all radar sets contain. More sophisticated methods of signal processing are also nearly always used in order to recover useful radar signals. The weak absorption of radio waves by the medium through which it passes is what enables radar sets to detect objects at relatively-long ranges—ranges at which other electromagnetic wavelengths, such as visible light, infrared light, and ultraviolet light, are too strongly attenuated. In particular, there are weather conditions under which radar works well regardless of the weather. Such things as fog, clouds, rain, falling snow, and sleet that block visible light are usually transparent to radio waves. Certain, specific radio frequencies that are absorbed or scattered by water vapour, raindrops, or atmospheric gases (especially oxygen) are avoided in designing radars except when detection of these is intended.

Finally, radar relies on its own transmissions, rather than light from the Sun or the Moon, or from electromagnetic waves emitted by the objects themselves, such as infrared wavelengths (heat). This process of directing artificial radio waves towards objects is called *illumination*, regardless of the fact that radio waves are completely invisible to the human eye or cameras.

Reflection

Electromagnetic waves reflect (scatter) from any large change in the dielectric constant or diamagnetic constants. This means that a solid object in air or a vacuum, or other significant change in atomic density between the object and what is surrounding it, will usually scatter radar (radio) waves. This is particularly true for electrically conductive materials, such as metal and carbon fibre, making radar particularly well suited to the detection of aircraft and ships. Radar absorbing material, containing resistive and sometimes magnetic substances, is used on military vehicles to reduce radar reflection. This is the radio equivalent of painting something a dark colour so that it cannot be seen through normal means.

Radar waves scatter in a variety of ways depending on the size (wavelength) of the radio wave and the shape of the target. If the wavelength is much shorter than the target's size, the wave will bounce off in a way similar to the way light is reflected by a mirror. If the wavelength is much longer than the size of the target, the target may not be visible due to poor reflection. Low Frequency radar technology is dependent on resonances for detection, but not identification, of targets. This is described by Rayleigh scattering, an effect that creates the Earth's blue sky and red sunsets. When the two length scales are comparable, there may be resonances. Early radars used very long wavelengths that were larger than the targets and received a vague signal, whereas some modern systems use shorter wavelengths (a few centimetres or shorter) that can image objects as small as a loaf of bread.

Short radio waves reflect from curves and corners, in a way similar to glint from a rounded piece of glass. The most reflective targets for short wavelengths have 90° angles between the reflective surfaces. A structure consisting of three flat surfaces meeting at a single corner, like the corner on a box, will always reflect waves entering its opening directly back at the source.

These so-called corner reflectors are commonly used as radar reflectors to make otherwise difficult-to-detect objects easier to detect, and are often found on boats in order to improve their detection in a rescue situation and to reduce collisions. For similar reasons, objects attempting to avoid detection will angle their surfaces in a way to eliminate inside corners and avoid surfaces and edges perpendicular to likely detection directions, which leads to "odd" looking stealth aircraft. These precautions do not completely eliminate reflection because of diffraction, especially at longer wavelengths. Half wavelength long wires or strips of conducting material, such as chaff, are very reflective but do not direct the scattered energy back toward the source. The extent to which an object reflects or scatters radio waves is called its radar cross section.

Doppler Effect

Ground-based radar systems used for detecting speeds rely on the Doppler effect. The apparent frequency (f) of the wave changes with the relative position of the target. The doppler equation is stated as follows for v_{obs} (the radial speed of the observer) and v_s (the radial speed of the target) and f_0 frequency of wave :

$$f = \frac{v + v_{obs}}{v - v_s} f_0$$

However, the change in phase of the return signal is often used instead of the change in frequency. It is to be noted that only the radial component of the speed is available. Hence when a target is moving at right angle to the radar beam, it has no velocity while one parallel to it has maximum recorded speed even if both might have the same real absolute motion.

Polarization

In the transmitted radar signal, the electric field is perpendicular to the direction of propagation, and this direction of the electric field is the polarization of the wave. Radars use horizontal, vertical, linear and circular polarization to detect different types of reflections. For example, circular polarization is used to minimize the interference caused by rain. Linear polarization returns usually indicate metal surfaces. Random polarization returns usually indicate a fractal surface, such as rocks or soil, and are used by navigation radars.

Limiting Factors

Beam Path and Range

The radar beam would follow a linear path in vacuum but it really follows a somewhat curved path in the atmosphere due to the variation of the refractive index of air. Even when the beam is emitted parallel to the ground, it will raise above it as the Earth curvature sink below the horizon. Furthermore, the signal is attenuated by the medium it crosses and the beam disperse as its not a perfect pencil shape.

The maximum range of a conventional radar can either be limited by a number of factors:

1. Line of sight, which depends on height above ground
2. The maximum non-ambiguous range (MUR) which is determined by the Pulse repetition frequency (PRF). Simply put, MUR is the distance the pulse could travel and return before the next pulse is emitted
3. Radar sensitivity and power of the return signal as computed in the radar equation. This includes factors such as environmentals and the size (or radar cross section) of the target.

Noise

Signal noise is an internal source of random variations in the signal, which is generated by all electronic components. Noise typically appears as random variations superimposed on the desired echo signal received in the radar receiver. The lower the power of the desired signal, the more difficult it is to discern it from the noise (similar to trying to hear a whisper while standing near a busy road). Noise figure is a measure of the noise produced by a receiver compared to an ideal receiver, and this needs to be minimized.

Noise is also generated by external sources, most importantly the natural thermal radiation of the background scene surrounding the target of interest. In modern radar systems, due to the high performance of their receivers, the internal noise is typically about equal to or lower than the external scene noise. An exception is if the radar is aimed upwards at clear sky, where the scene is so "cold" that it generates very little thermal noise.

There will be also flicker noise due to electrons transit, but depending on 1/f, will be much lower than thermal noise when the frequency is high. Hence, in pulse radar, the system will be always heterodyne.

Interference

Radar systems must overcome unwanted signals in order to focus only on the actual targets of interest. These unwanted signals may originate from internal and external sources, both passive and active. The ability of the radar system to overcome these unwanted signals defines its signal-to-noise ratio (SNR). SNR is defined as the ratio of a signal power to the noise power within the desired signal.

In less technical terms, SNR compares the level of a desired signal (such as targets) to the level of background noise. The higher a system's SNR, the better it is in isolating actual targets from the surrounding noise signals.

Clutter

Clutter refers to radio frequency (RF) echoes returned from targets which are uninteresting to the radar operators. Such targets include natural objects such as ground, sea, precipitation (such as rain, snow or hail), sand storms, animals (especially birds), atmospheric turbulence, and other atmospheric effects, such as ionosphere reflections, meteor trails, and three body scatter spike. Clutter may also be returned from man-made objects such as buildings and, intentionally, by radar countermeasures such as chaff.

Some clutter may also be caused by a long radar waveguide between the radar transceiver and the antenna. In a typical plan position indicator (PPI) radar with a rotating antenna, this will usually be seen as a "sun" or "sunburst" in the centre of the display as the receiver responds to echoes from dust particles and misguided RF in the waveguide. Adjusting the timing between when the transmitter sends a pulse and when the receiver stage is enabled will generally reduce the sunburst without affecting the accuracy of the range, since most sunburst is caused by a diffused transmit pulse reflected before it leaves the antenna.

While some clutter sources may be undesirable for some radar applications (such as storm clouds for air-defence radars), they may be desirable for others (meteorological radars in this example). Clutter is considered a passive interference source, since it only appears in response to radar signals sent by the radar. There are several methods of detecting and neutralizing clutter. Many of these methods rely on the fact that clutter tends to appear static between radar scans. Therefore, when comparing subsequent scans echoes, desirable targets will appear to move and all stationary echoes can be eliminated. Sea clutter can be reduced by using horizontal polarization, while rain is reduced with circular polarization (note that meteorological radars wish for the opposite effect, therefore using linear polarization the better to detect precipitation). Other methods attempt to increase the signal-to-clutter ratio.

Constant False Alarm Rate (CFAR, a form of Automatic Gain Control, or AGC) is a method relying on the fact that clutter returns far outnumber echoes from targets of interest. The receiver's gain is automatically adjusted to maintain a constant level of overall visible clutter. While this does not help detect targets masked by stronger surrounding clutter, it does help to distinguish strong target sources. In the past, radar AGC was electronically controlled and affected the gain of the entire radar receiver. As radars evolved, AGC became computer-software controlled, and affected the gain with greater granularity, in specific detection cells.

Clutter may also originate from multipath echoes from valid targets due to ground reflection, atmospheric ducting or ionospheric reflection/refraction (e.g. Anomalous propagation). This clutter type is especially bothersome, since it appears to move and behave like other normal (point) targets of interest, thereby creating a ghost. In a typical scenario, an aircraft echo is multipath-reflected from the ground below, appearing to the receiver as an identical target below the correct one. The radar may try to unify the targets, reporting the target at an incorrect height, or—worse—eliminating it on the basis of jitter or a physical impossibility. These problems can be overcome by incorporating a ground map of the radar's surroundings and eliminating all echoes which appear to originate below ground or above a certain height. In newer Air Traffic Control (ATC) radar equipment, algorithms are used to identify the false targets by comparing the current pulse returns, to those adjacent, as well as calculating return improbabilities due to calculated height, distance, and radar timing.

Jamming

Radar jamming refers to radio frequency signals originating from sources outside the radar, transmitting in the radar's frequency and thereby masking targets of interest. Jamming may be intentional, as with an electronic warfare (EW) tactic, or unintentional, as with friendly forces operating equipment that transmits using the same frequency range. Jamming is considered an active interference source, since it is initiated by elements outside the radar and in general unrelated to the radar signals.

Jamming is problematic to radar since the jamming signal only needs to travel one-way (from the jammer to the radar receiver) whereas the radar echoes travel two-ways (radar-target-radar) and are therefore significantly reduced in power by the time they return to the radar receiver. Jammers therefore can be much less powerful than their jammed radars and still effectively mask targets along the line of sight from the jammer to the radar (*Mainlobe Jamming*). Jammers have an added effect of affecting radars along other lines of sight, due to the radar receiver's sidelobes (*Sidelobe Jamming*).

Mainlobe jamming can generally only be reduced by narrowing the mainlobe solid angle, and can never fully be eliminated when directly facing a jammer which uses the same frequency and polarization as the radar. Sidelobe jamming can be overcome by reducing receiving sidelobes in the radar antenna design and by using an omnidirectional antenna to detect and disregard non-mainlobe signals. Other anti-jamming techniques are frequency hopping and polarization.

Interference has recently become a problem for C-band (5.66 GHz) meteorological radars with the proliferation of 5.4 GHz band WiFi equipment.

Chapter 2

Data Acquisition Techniques

The basis for multispectral collection and analysis is that of examined areas or objects that reflect or emit radiation that stand out from surrounding areas.

Applications

Radar

Radar is an object-detection system which uses radio waves to determine the range, altitude, direction, or speed of objects. It can be used to detect aircraft, ships, spacecraft, guided missiles, motor vehicles, weather formations, and terrain. The radar dish or antenna transmits pulses of radio waves or microwaves which bounce off any object in their path. The object returns a tiny part of the wave's energy to a dish or antenna which is usually located at the same site as the transmitter. Radar was developed in secret in nations across the world during World War II. The term RADAR was coined in 1940 by the United States Navy as an acronym for radio detection and ranging. The term radar has since entered English and other languages as the common noun radar, losing all capitalization.

The modern uses of radar are highly diverse, including air traffic control, radar astronomy, air-defense systems, antimissile systems; marine radars to locate landmarks and other ships; aircraft anticollision systems; ocean surveillance systems, outer space surveillance and rendezvous systems; meteorological precipitation monitoring; altimetry and flight control systems; guided missile target locating systems; and ground-penetrating radar for geological observations. High tech radar systems are associated with digital signal processing and are capable of extracting objects from very high noise levels.

Other systems similar to radar have been used in other parts of the electromagnetic spectrum. One example is "lidar", which uses visible light from lasers rather than radio waves.

Figure : Radar-dish Antenna

As early as 1886, Heinrich Hertz showed that radio waves could be reflected from solid objects. In 1895 Alexander Popov, a physics instructor at the Imperial Russian Navy school in Kronstadt, developed an apparatus using a coherer tube for detecting distant lightning strikes. The next year, he added a spark-gap transmitter. In 1897, while testing this in communicating between two ships in the Baltic Sea, he took note of an interference beat caused by the passage of a third vessel. In his report, Popov wrote that this phenomenon might be used for detecting objects, but he did nothing more with this observation.

The German Christian Huelsmeyer was the first to use radio waves to detect "the presence of distant metallic objects". In 1904 he demonstrated the feasibility of detecting a ship in dense fog but not its distance. He obtained a patent for his detection device in April 1904 and later a patent for a related amendment for determining the distance to the ship. He also got a British patent on September 23, 1904 for the first full radar application, which he called telemobiloscope.

In August 1917 Nikola Tesla outlined a concept for primitive radar units. He stated:

> *"...by their [standing electromagnetic waves] use we may produce at will, from a sending station, an electrical effect in any particular region of the globe; [with which] we may determine the relative position or course of a moving object, such as a vessel at sea, the distance traversed by the same, or its speed."*

In 1922 A. Hoyt Taylor and Leo C. Young, researchers working with the U.S. Navy, discovered that when radio waves were broadcast at 60 MHz it was possible to determine the range and bearing of nearby ships in the Potomac River. Despite Taylor's suggestion that this method could be used in darkness and low visibility, the Navy did not immediately continue the work. Serious investigation began eight years later after the discovery that radar could be used to track airplanes.

Before the Second World War, researchers in France, Germany, Italy, Japan, the Netherlands, the Soviet Union, the United Kingdom, and the United States, independently and in great secrecy, developed technologies that led to the modern version of radar. Australia, Canada, New Zealand, and South Africa followed prewar Great Britain, and Hungary had similar developments during the war.

In 1934 the Frenchman Émile Girardeau stated he was building an obstacle-locating radio apparatus "conceived according to the principles stated by Tesla" and obtained a patent for a working system, a part of which was installed on the Normandie liner in 1935.

During the same year, the Soviet military engineer P.K.Oschepkov, in collaboration with Leningrad Electrophysical Institute, produced an experimental apparatus, RAPID, capable of detecting an aircraft within 3 km of a receiver. The French and Soviet systems, however, had continuous-wave operation and could not give the full performance that was ultimately at the centre of modern radar.

Full radar evolved as a pulsed system, and the first such elementary apparatus was demonstrated in December 1934 by

American Robert M. Page, working at the Naval Research Laboratory. The following year, the United States Army successfully tested a primitive surface to surface radar to aim coastal battery search lights at night.

This was followed by a pulsed system demonstrated in May 1935 by Rudolf Kuhnhold and the firm GEMA in Germany and then one in June 1935 by an Air Ministry team led by Robert A. Watson Watt in Great Britain. Later, in 1943, Page greatly improved radar with the monopulse technique that was used for many years in most radar applications.

The British were the first to fully exploit radar as a defence against aircraft attack. This was spurred on by fears that the Germans were developing death rays. The Air Ministry asked British scientists in 1934 to investigate the possibility of propagating electromagnetic energy and the likely effect. Following a study, they concluded that a death ray was impractical but that detection of aircraft appeared feasible. Robert Watson Watt's team demonstrated to his superiors the capabilities of a working prototype and then patented the device. It served as the basis for the Chain Home network of radars to defend Great Britain.

In April 1940, Popular Science showed an example of a radar unit using the Watson-Watt patent in an article on air defence, but not knowing that the U.S. Army and U.S. Navy were working on radars with the same principle, stated under the illustration, "This is not U.S. Army equipment." Also, in late 1941 Popular Mechanics had an article in which a U.S. scientist conjectured what he believed the British early warning system on the English east coast most likely looked like and was very close to what it actually was and how it worked in principle.

The war precipitated research to find better resolution, more portability, and more features for radar, including complementary navigation systems like Oboe used by the RAF's Pathfinder.

The information provided by radar includes the bearing and range (and therefore position) of the object from the radar scanner. It is thus used in many different fields where the need for such positioning is crucial. The first use of radar was for military purposes: to locate air, ground and sea targets. This evolved in the civilian field into applications for aircraft, ships, and roads.

In aviation, aircraft are equipped with radar devices that warn of obstacles in or approaching their path and give accurate altitude readings. The first commercial device fitted to aircraft was a 1938 Bell Lab unit on some United Air Lines aircraft. They can land in fog at airports equipped with radar-assisted ground-controlled approach systems, in which the plane's flight is observed on radar screens while operators radio landing directions to the pilot.

Marine radars are used to measure the bearing and distance of ships to prevent collision with other ships, to navigate and to fix their position at sea when within range of shore or other fixed references such as islands, buoys, and lightships. In port or in harbour, vessel traffic service radar systems are used to monitor and regulate ship movements in busy waters. Police forces use radar guns to monitor vehicle speeds on the roads.

Meteorologists use radar to monitor precipitation. It has become the primary tool for short-term weather forecasting and to watch for severe weather such as thunderstorms, tornadoes, winter storms, precipitation types, etc. Geologists use specialised ground-penetrating radars to map the composition of the Earth's crust.

Principles: A radar system has a transmitter that emits radio waves called radar signals in predetermined directions. When these come into contact with an object they are usually reflected and/or scattered in many directions. Radar signals are reflected especially well by materials of considerable electrical conductivity—especially by most metals, by seawater, by wet land, and by wetlands. Some of these make the use of radar altimeters possible. The radar signals that are reflected back towards the transmitter are the desirable ones that make radar work. If the object is moving either closer or farther away, there is a slight change in the frequency of the radio waves, caused by the Doppler effect.

Radar receivers are usually, but not always, in the same location as the transmitter. Although the reflected radar signals captured by the receiving antenna are usually very weak, these signals can be strengthened by the electronic amplifiers. More sophisticated methods of signal processing are also used in order to recover useful radar signals.

The weak absorption of radio waves by the medium through which it passes is what enables radar sets to detect objects at relatively long ranges—ranges at which other electromagnetic wavelengths, such as visible light, infrared light, and ultraviolet light, are too strongly attenuated. Such things as fog, clouds, rain, falling snow, and sleet that block visible light are usually transparent to radio waves. Certain radio frequencies that are absorbed or scattered by water vapour, raindrops, or atmospheric gases (especially oxygen) are avoided in designing radars except when detection of these is intended.

Radar relies on its own transmissions rather than light from the Sun or the Moon, or from electromagnetic waves emitted by the objects themselves, such as infrared wavelengths (heat). This process of directing artificial radio waves towards objects is called illumination, regardless of the fact that radio waves are invisible to the human eye or cameras.

Electromagnetic waves reflect (scatter) from any large change in the dielectric constant or diamagnetic constants. This means that a solid object in air or a vacuum, or other significant change in atomic density between the object and what is surrounding it, will usually scatter radar (radio) waves.

This is particularly true for electrically conductive materials, such as metal and carbon fibre, making radar well suited to the detection of aircraft and ships. Radar absorbing material, containing resistive and sometimes magnetic substances, is used on military vehicles to reduce radar reflection. This is the radio equivalent of painting something a dark colour so that it cannot be seen through normal means.

Radar waves scatter in a variety of ways depending on the size (wavelength) of the radio wave and the shape of the target. If the wavelength is much shorter than the target's size, the wave will bounce off in a way similar to the way light is reflected by a mirror. If the wavelength is much longer than the size of the target, the target may not be visible because of poor reflection. Low Frequency radar technology is dependent on resonances for detection, but not identification, of targets. This is described by Rayleigh scattering, an effect that creates the Earth's blue sky and red sunsets. When the two length scales are comparable, there may be resonances. Early radars used very long wavelengths that were larger than the targets and received a vague signal, whereas some modern systems use shorter wavelengths (a few centimetres or shorter) that can image objects as small as a loaf of bread.

Short radio waves reflect from curves and corners, in a way similar to glint from a rounded piece of glass. The most reflective targets for short wavelengths have 90° angles between the reflective surfaces. A structure consisting of three flat surfaces meeting at a single corner, like the corner on a box, will reflect waves entering its opening directly back at the source. These so-called corner reflectors are commonly used as radar reflectors to make otherwise difficult-to-detect objects easier to detect and are often found on boats in order to improve their detection in a rescue situation and to reduce collisions. For similar reasons, objects attempting to avoid detection will angle their surfaces in a way to eliminate inside corners and avoid surfaces and edges perpendicular to likely detection directions, which leads to "odd" looking stealth aircraft. These precautions do not completely eliminate reflection because of diffraction, especially at longer wavelengths. Half wavelength long wires or strips of conducting material, such as chaff, are very reflective but do not direct the scattered energy back toward the source. The extent to which an object reflects or scatters radio waves is called its radar cross section.

LIDAR

Figure : *A FASOR used at the Starfire Optical Range for LIDAR and laser guide star experiments is tuned to the sodium D2a line and used to excite sodium atoms in the upper atmosphere.*

LIDAR (Light Detection And Ranging, also LADAR) is an optical remote sensing technology that can measure the distance to, or other properties of a target by illuminating the target with light, often using pulses from a laser. LIDAR technology has application in geomatics, archaeology, geography, geology, geomorphology, seismology, forestry, remote sensing and atmospheric physics, as well as in airborne laser swath mapping (ALSM), laser altimetry and LIDAR contour mapping.

The acronym LADAR (Laser Detection and Ranging) is often used in military contexts. The term "laser radar" is sometimes used, even though LIDAR does not employ microwaves or radio waves and therefore is not radar in the strict sense of the word.

LIDAR uses ultraviolet, visible, or near infrared light to image objects and can be used with a wide range of targets, including non-metallic objects, rocks, rain, chemical compounds, aerosols, clouds and even single molecules. A narrow laser beam can be used to map physical features with very high resolution.

LIDAR has been used extensively for atmospheric research and meteorology. Downward-looking LIDAR instruments fitted to aircraft and satellites are used for surveying and mapping – a recent example being the NASA Experimental Advanced Research Lidar. In addition LIDAR has been identified by NASA as a key technology for enabling autonomous precision safe landing of future robotic and crewed lunar landing vehicles.

Figure : This LIDAR-equipped mobile robot uses its LIDAR to construct a map and avoid obstacles.

Wavelengths in a range from about 10 micrometers to the UV (ca. 250 nm) are used to suit the target. Typically light is reflected via backscattering. Different types of scattering are used for different LIDAR applications; most common are Rayleigh scattering, Mie scattering and Raman scattering, as well as fluorescence. Based on different kinds of backscattering, the LIDAR can be accordingly called Rayleigh LiDAR, Mie LiDAR, Raman LiDAR and Na/Fe/K Fluorescence LIDAR and so on. Suitable combinations of wavelengths can allow for remote mapping of atmospheric contents by looking for wavelength-dependent changes in the intensity of the returned signal.

In general there are two kinds of lidar detection schema: "incoherent" or direct energy detection (which is principally an amplitude measurement) and Coherent detection (which is best for doppler, or phase sensitive measurements). Coherent systems generally use Optical heterodyne detection which being more sensitive than direct detection allows them to operate a much lower power but at the expense of more complex transceiver requirements.

In both coherent and incoherent LIDAR, there are two types of pulse models: micropulse lidar systems and high energy systems. Micropulse systems have developed as a result of the ever increasing amount of computer power available combined with advances in laser technology. They use considerably less energy in the laser, typically on the order of one microjoule, and are often "eye-safe," meaning they can be used without safety precautions. High-power systems are common in atmospheric research, where they are widely used for measuring many atmospheric parameters: the height, layering and densities of clouds, cloud particle properties (extinction coefficient, backscatter coefficient, depolarization), temperature, pressure, wind, humidity, trace gas concentration (ozone, methane, nitrous oxide, etc.).

There are several major components to a LIDAR system:

- Laser — 600–1000 nm lasers are most common for non-scientific applications. They are inexpensive, but since they can be focused and easily absorbed by the eye, the maximum power is limited by the need to make them eye-safe. Eye-safety is often a requirement for most applications. A common alternative, 1550 nm lasers, are eye-safe at much higher power levels since this wavelength is not focused by the eye, but the detector technology is less advanced and so these wavelengths are generally used at longer ranges and lower accuracies. They are also used for military applications as 1550 nm is not visible in night vision goggles, unlike the shorter 1000 nm infrared laser. Airborne topographic mapping lidars generally use 1064 nm diode pumped YAG lasers, while bathymetric systems generally use 532 nm frequency doubled diode pumped YAG lasers because 532 nm penetrates water with much less attenuation than does 1064 nm. Laser settings include the laser repetition rate (which controls the data collection speed). Pulse length is generally an attribute of the laser cavity length, the number of passes required through the gain material (YAG, YLF, etc.), and Q-switch speed. Better target resolution is achieved with shorter pulses, provided the LIDAR receiver detectors and electronics have sufficient bandwidth.

- Scanner and optics — How fast images can be developed is also affected by the speed at which they are scanned. There are several options to scan the azimuth and elevation, including dual oscillating plane mirrors, a combination with a polygon mirror, a dual axis scanner (see Laser scanning). Optic choices affect the angular resolution and range that can be detected. A hole mirror or a beam splitter are options to collect a return signal.
- Photodetector and receiver electronics — Two main photodetector technologies are used in lidars: solid state photodetectors, such as silicon avalanche photodiodes, or photomultipliers. The sensitivity of the receiver is another parameter that has to be balanced in a LIDAR design.
- Position and navigation systems — LIDAR sensors that are mounted on mobile platforms such as airplanes or satellites require instrumentation to determine the absolute position and orientation of the sensor. Such devices generally include a Global Positioning System receiver and an Inertial Measurement Unit (IMU).
- 3D imaging can be achieved using both scanning and non-scanning systems. "3D gated viewing laser radar" is a non-scanning laser ranging system that applies a pulsed laser and a fast gated camera.
- Imaging LIDAR can also be performed using arrays of high speed detectors and modulation sensitive detectors arrays typically built on single chips using CMOS and hybrid CMOS/ CCD fabrication techniques. In these devices each pixel

performs some local processing such as demodulation or gating at high speed down converting the signals to video rate so that the array may be read like a camera. Using this technique many thousands of pixels / channels may be acquired simultaneously. High resolution 3D LIDAR cameras use homodyne detection with an electronic CCD or CMOS shutter.

A coherent Imaging LIDAR uses Synthetic array heterodyne detection to enables a staring single element receiver to act as though it were an imaging array.

Radiometer

Figure : *An example of a Crookes radiometer. The vanes rotate when exposed to light, with faster rotation for more intense light, providing a quantitative measurement of electromagnetic radiation intensity.*

A radiometer is a device for measuring the radiant flux (power) of electromagnetic radiation. Generally, the term radiometer denotes an infrared radiation detector, yet it also includes detectors operating on any electromagnetic wavelength.

A common example is the Crookes radiometer, an early-model device wherein a rotor (having vanes which are dark on one side, and light on the other) in a partial vacuum spins when exposed to light. A common myth (one originally held even by Crookes) is that the momentum of the absorbed light on the black faces makes the radiometer operate. If this were true however, the radiometer would spin away from the non-black faces, since the photons bouncing off those faces impart even more momentum than the photons absorbed on the black faces. Follow the link below for an in-depth explanation of the principles behind a Crookes radiometer. The Nichols radiometer operates on a different principle and is more sensitive than the Crookes type.

A microwave radiometer operates in the microwave wavelengths. The radiometer contains argon gas to enable it to rotate.

The MEMS radiometer, invented by Patrick Jankowiak, can operate on the principles of Nichols or Crooke and can operate over a wide spectrum of wavelength and particle energy levels.

Photometer

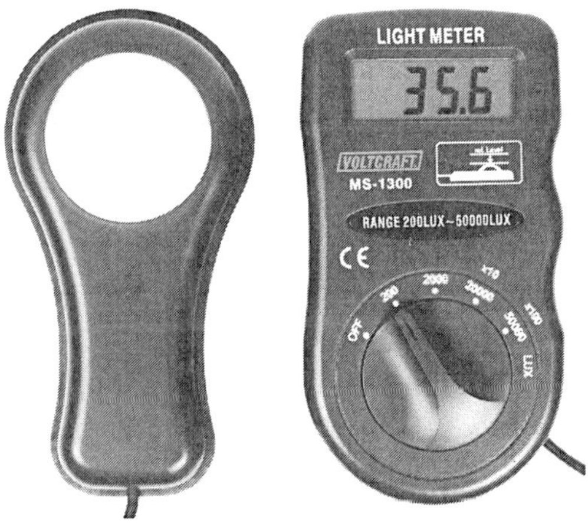

Figure: *A photometer*

In its widest sense, a photometer is an instrument for measuring light intensity or optical properties of solutions or surfaces. Photometers are used to measure:

- Illuminance
- Irradiance
- Light absorption
- Scattering of light
- Reflection of light
- Fluorescence
- Phosphorescence
- Luminescence

Before electronic light sensitive elements were developed, photometry was done by estimation by the eye. The relative luminous flux of a source was compared with a standard source. The photometer is placed such that the illuminance from the source being investigated is equal to that of the standard source as equal illuminance can be judged by the eye.

The relative luminous fluxes can then be calculated as the illuminance decreases proportionally to the inverse square of distance. A standard example of such a photometer consists of a piece of paper with an oil spot on it that makes the paper there slightly more transparent. When the spot is not visible from either side, the illuminance from the two sides is equal.

Most photometers detect the light with photoresistors, photodiodes or photomultipliers. To analyse the light, the photometer may measure the light after it has passed through a filter or through a monochromator for determination at defined wavelengths or for analysis of the spectral distribution of the light.

Some photometers measure light by counting individual photons rather than incoming flux. The operating principles are the same but the results are given in units such as photons/cm2 or photons ·cm-2 ·sr-1 rather than W/cm2 or W ·cm-2 ·sr-1.

Due to their individual photon counting nature, these instruments are limited to observations where the irradiance is low. The irradiance is limited by the time resolution of its associated detector readout electronics. With current technology this is in the megahertz range. The maximum irradiance is also limited by the throughput and gain parameters of the detector itself.

The light sensing element in photon counting devices in NIR, visible and ultrviolet wavelengths is a photomultiplier to achieve sufficient sensitivity.

In airborne and space-based remote sensing such photon counters are used at the upper reaches of the electromagnetic spectrum such as the X-ray to far ultraviolet. This is usually due to the lower radiant intensity of the objects being measured as well as the difficulty of measuring light at higher energies using its particle-like nature as compared to the wavelike nature of light at lower frequencies. Conversely, radiometers are typically used for remote sensing from the visible, infrared though radio frequency range.

Photometers are used to determine the correct exposure in photography. In modern cameras, the photometer is usually built in. As the illumination of different parts of the picture varies, advanced photometers measure the light intensity in different parts of the potential picture and use an algorithm to determine the most suitable exposure for the final picture, adapting the algorithm to the type of picture intended(see Metering mode). Historically, a photometer was separate from the camera. The advanced photometers then could be used either to measure the light from the potential picture as a whole, to measure from elements of the picture to ascertain that the most important parts of the picture are optimally exposed, or to measure the incident light to the scene with an integrating adapter.

A reflectance photometer measures the reflectance of a surface as a function of wavelength. The surface is illuminated with white light, and the reflected light is measured after passing through a monochromator. This type of measurement has mainly practical applications, for instance in the paint industry to characterize the colour of a surface objectively.

These are optical instruments for measurement of the absorption of light of a given wavelength (or a given range of wavelengths) of coloured substances in solution. From the light absorption, Beer's law makes it possible to calculate the concentration of the coloured substance in the solution. Due to its wide range of application and its reliability and robustness, the photometer has become one of the principal instruments in biochemistry and analytical chemistry. Absorption photometers for work in aqueous solution work in the ultraviolet and visible ranges, from wavelength around 240 nm up to 750 nm.

The principle of spectrophotometers and filter photometers is that (as far as possible) monochromatic light is allowed to pass through a container (cell) with optically flat windows containing the solution. It then reaches a light detector, that measures the intensity of the light compared to the intensity after passing through an identical cell with the same solvent but without the coloured substance. From the ratio between the light intensities, knowing the capacity of the coloured substance to absorb light (the absorbancy of the coloured substance, or the photon cross section area of the molecules of the coloured substance at a given wavelength), it is possible to calculate the concentration of the substance using Beer's law.

Two types of photometers are used: spectrophotometer and filter photometer. In spectrophotometers a monochromator (with prism or with grating) is used to obtain monochromatic light of one defined wavelength. In filter photometers, optical filters are used to give the monochromatic light. Spectrophotometers can thus easily be set to measure the absorbance at different wavelengths, and they can also be used to scan the spectrum of the absorbing substance. They are in this way more flexible than filter photometers, also give a higher optical purity of the analysing light, and therefore they are preferably used for research purposes. Filter photometers are cheaper, robuster and easier to use and therefore they are used for routine analysis. Photometers for microtiter plates are filter photometers.

Spectrophotometry in infrared light is mainly used to study structure of substances, as given groups give absorption at defined wavelengths. Measurement in aqueous solution is generally not possible, as water absorbs infrared light strongly in some wavelength ranges. Therefore, infrared spectroscopy is either performed in the gaseous phase (for volatile substances) or with the substances pressed into tablets together with salts that are transparent in the infrared range. Potassium bromide (KBr) is commonly used for this purpose. The substance to be tested is thoroughly mixed with specially purified KBr and pressed into a transparent tablet, that is placed in the beam of light. The analysis of the wavelength dependence is generally not done using a monochromator as it is in UV-Vis, but with the use of an interferometer.

The interference pattern can be analysed using a Fourier transform algorithm. In this way, the whole wavelength range can be analysed simultaneously, saving time, and an interferometer is also less expensive than a monochromator. The light absorbed in the infrared region does not correspond to electronic excitation of the substance studied, but rather to different kinds of vibrational excitation. The vibrational excitations are characteristic of different groups in a molecule, that can in this way be identified. The infrared spectrum typically has very narrow absorption lines, which makes them unsuited for quantitative analysis but gives very detailed information about the molecules. The frequencies of the different modes of vibration varies with isotope, and therefore different isotopes give different peaks. This makes it possible also to study the isotopic composition of a sample with infrared spectrophotometry.

Atomic absorption photometers are photometers that measure the light from a very hot flame. The solution to be analysed is injected into the flame at a constant, known rate. Metals in the solution are present in atomic form in the flame. The monochromatic light in this type of photometer is generated by a discharge lamp where the discharge takes place in a gas with the metal to be determined. The discharge then emits light with wavelengths corresponding to the spectral lines of the metal. A filter may be used to isolate one of the main spectral lines of the metal to be analysed. The light is absorbed by the metal in the flame, and the absorption is used to determine the concentration of the metal in the original solution.

Aerial Photography

Aerial photography is the taking of photographs of the ground from an elevated position. The term usually refers to images in which the camera is not supported by a ground-based structure. Cameras may be hand held or mounted, and photographs may be taken by a photographer, triggered remotely or triggered automatically. Platforms for aerial photography include fixed-wing aircraft, helicopters, balloons, blimps and dirigibles, rockets, kites, poles, parachutes, and vehicle mounted poles . Aerial photography should not be confused with Air-to-Air Photography, when aircraft serve both as a photo platform and subject.

Figure : *Aerial photography of Machu Picchu, Peru*

Aerial photography was first practiced by the French photographer and balloonist Gaspard-Félix Tournachon, known as "Nadar", in 1858 over Paris, France. The first use of a motion picture camera mounted to a heavier-than-air aircraft took place on April 24, 1909 over Rome in the 3:28 silent film short, Wilbur Wright und seine Flugmaschine.

The first special semiautomatic aerial camera was designed in 1911 by Russian military engineer — Colonel Potte V. F. This aerial camera was used during World War I.

The use of aerial photography for military purposes was expanded during World War I by many other aviators such as Fred Zinn. One of the first notable battles was that of Neuve Chapelle.

Figure: *New York City Aerials*

Aerial mapping came into use on the battlefronts during World War I. In January 1918, General Allenby used five Australian pilots from No. 1 Squadron AFC to photograph a 624 square miles (1,620 km2) area in Palestine as an aid to correcting and improving maps of the Turkish front. Lieutenants Leonard Taplin, Allan Runciman Brown, H. L. Fraser, Edward Patrick Kenny, and L. W. Rogers photographed a block of land stretching from the Turkish front lines 32 miles (51 km) deep into their rear areas. Beginning 5 January, they flew with a fighter escort to ward off enemy fighters. Using Royal Aircraft Factory BE.12 and Martinsyde airplanes, they not only overcame enemy air attacks, but also bucked 65 mile per hour winds, antiaircraft fire, and malfunctioning equipment to complete their task circa 19 January 1918.

One of the most successful pioneers of the commercial use of aerial photography was by Sherman Fairchild who started his own aircraft firm Fairchild Aircraft to develop and build specialized aircraft for high altitude aerial survey missions. One Fairchild aerial survey aircraft in 1935 carried unit that combined two synchronized cameras, and each camera having five six inch lenses with a ten inch lenses and took photos from 23,000 feet. Each photo cover two hundred and twenty five square miles. One of its first government contracts was an aerial survey of New Mexico to study soil erosion. A year later, Fairchild introduced a better high altitude camera with nine-lens in one unit that could take a photo of 600 square miles with each exposure from 30,000 feet.

With the advent of inexpensive digital cameras, many people now take candid photographs from commercial aircraft and increasingly from general aviation aircraft on private pleasure flights.

Aerial Photography Platforms: Advances in radio controlled models have made it possible for model aircraft to conduct low-altitude aerial photography. This has benefited real-estate advertising, where commercial and residential properties are the photographic subject. Full-size, manned aircraft are prohibited from low flights above populated locations. Small scale model aircraft offer increased photographic access to these previously restricted areas. Miniature vehicles do not replace full size aircraft, as full size aircraft are capable of longer flight times, higher altitudes, and greater equipment payloads. They are, however, useful in any situation in which a full-scale aircraft would be dangerous to operate. Examples would include the inspection of transformers atop power transmission lines and slow, low-level flight over agricultural fields, both of which can be accomplished by a large-scale radio controlled helicopter. Professional-grade, gyroscopically stabilized camera platforms are available for use under such a model; a large model helicopter with a 26cc gasoline engine can hoist a payload of approximately seven kilograms (15 lbs)

Recent (2006) FAA regulations grounding all commercial RC model flights have been upgraded to require formal FAA certification before permission to fly at any altitude in USA.

Because anything capable of being viewed from a public space is considered outside the realm of privacy in the United States, aerial photography may legally document features and occurrences on private property.

Geodetic

Overhead geodetic collection was first used in aerial submarine detection and gravitational data used in military maps. This data revealed minute perturbations in the Earth's gravitational field (geodesy) that may be used to determine changes in the mass distribution of the Earth, which in turn may be used for geological studies.

Acoustic and Near-acoustic

- Sonar: *passive sonar*, listening for the sound made by another object (a vessel, a whale etc); *active sonar*, emitting pulses of sounds and listening for echoes, used for detecting, ranging and measurements of underwater objects and terrain.
- Seismograms taken at different locations can locate and measure earthquakes (after they occur) by comparing the relative intensity and precise timing.

To coordinate a series of large-scale observations, most sensing systems depend on the following: platform location, what time it is, and the rotation and orientation of the sensor. High-end instruments now often use positional information from satellite navigation systems. The rotation and orientation is often provided within a degree or two with electronic compasses. Compasses can measure not just azimuth (i. e. degrees to magnetic north), but also altitude (degrees above the horizon), since the magnetic field curves into the Earth at different angles at different latitudes. More exact orientations require gyroscopic-aided orientation, periodically realigned by different methods including navigation from stars or known benchmarks.

Resolution impacts collection and is best explained with the following relationship: less resolution= less detail & larger coverage, More resolution= more detail, less coverage. The skilled management of collection results in cost-effective collection and avoid situations such as the use of multiple high resolution data which tends to clog transmission and storage infrastructure.

Data Processing

Generally speaking, remote sensing works on the principle of the *inverse problem*. While the object or phenomenon of interest (the state) may not be directly measured, there exists some other variable that can be detected and measured (the observation), which may be related to the object of interest through the use of a data-derived computer model. The common analogy given to describe this is trying to determine the type of animal from its footprints. For example, while it is impossible to directly measure temperatures in the upper atmosphere, it is possible to measure the spectral emissions from a known chemical species (such as carbon dioxide) in that region. The frequency of the emission may then be related to the temperature in that region via various thermodynamic relations.

The quality of remote sensing data consists of its spatial, spectral, radiometric and temporal resolutions.

Spatial Resolution: The size of a pixel that is recorded in a raster image typically pixels may correspond to square areas ranging in side length from 1 to 1,000 metres (3.3 to 3,300 ft).

Spectral Resolution: The wavelength width of the different frequency bands recorded – usually, this is related to the number of frequency bands recorded by the platform. Current Landsat collection is that of seven bands, including several in the infrared spectrum, ranging from a spectral resolution of 0.07 to 2.1 μm. The Hyperion sensor on Earth Observing-1 resolves 220 bands from 0.4 to 2.5 ìm, with a spectral resolution of 0.10 to 0.11 ìm per band.

Radiometric Resolution: The number of different intensities of radiation the sensor is able to distinguish. Typically, this ranges from 8 to 14 bits, corresponding to 256 levels of the gray scale and up to 16,384 intensities or "shades" of colour, in each band. It also depends on the instrument noise.

Temporal Resolution: The frequency of flyovers by the satellite or plane, and is only relevant in time-series studies or those requiring an averaged or mosaic image as in deforesting monitoring. This was first used by the intelligence community where repeated coverage revealed changes in infrastructure, the deployment of units or the modification/introduction of equipment. Cloud cover over a given area or object makes it necessary to repeat the collection of said location.

In order to create sensor-based maps, most remote sensing systems expect to extrapolate sensor data in relation to a reference point including distances between known points on the ground. This depends on the type of sensor used.

For example, in conventional photographs, distances are accurate in the centre of the image, with the distortion of measurements increasing the farther you get from the centre. Another factor is that of the platen against which the film is pressed can cause severe errors when photographs are used to measure ground distances. The step in which this problem is resolved is called georeferencing, and involves computer-aided matching up of points in the image (typically 30 or more points per image) which is extrapolated with the use of an established benchmark, "warping" the image to produce accurate spatial data. As of the early 1990s, most satellite images are sold fully georeferenced. In addition, images may need to be radiometrically and atmospherically corrected.

Radiometric Correction: gives a scale to the pixel values, e. g. the monochromatic scale of 0 to 255 will be converted to actual radiance values.

Atmospheric Correction: eliminates atmospheric haze by rescaling each frequency band so that its minimum value (usually realised in water bodies) corresponds to a pixel value of 0. The digitising of data also make possible to manipulate the data by changing gray-scale values. Interpretation is the critical process of making sense of the data.

The first application was that of aerial photographic collection which used the following process; spatial measurement through the use of a light table in both conventional single or stereographic coverage, added skills such as the use of photogrammetry, the use of photomosaics, repeat coverage, Making use of objects' known dimensions in order to detect modifications. Image Analysis is the recently developed automated computer-aided application which is in increasing use.

Object-Based Image Analysis (OBIA) is a sub-discipline of GIScience devoted to partitioning remote sensing (RS) imagery into meaningful image-objects, and assessing their characteristics through spatial, spectral and temporal scale.

Old data from remote sensing is often valuable because it may provide the only long-term data for a large extent of geography. At the same time, the data is often complex to interpret, and bulky to store. Modern systems tend to store the data digitally, often with lossless compression.

The difficulty with this approach is that the data is fragile, the format may be archaic, and the data may be easy to falsify. One of the best systems for archiving data series is as computer-generated machine-readable ultrafiche, usually in typefonts such as OCR-B, or as digitised half-tone images. Ultrafiches survive well in standard libraries, with lifetimes of several centuries. They can be created, copied, filed and retrieved by automated systems. They are about as compact as archival magnetic media, and yet can be read by human beings with minimal, standardised equipment.

Level Description: Reconstructed, unprocessed instrument and payload data at full resolution, with any and all communications artifacts (e. g., synchronization frames, communications headers, duplicate data) removed.

Reconstructed, unprocessed instrument data at full resolution, time-referenced, and annotated with ancillary information, including radiometric and geometric calibration coefficients and georeferencing parameters (e. g., platform ephemeris) computed and appended but not applied to the Level 0 data (or if applied, in a manner that level 0 is fully recoverable from level 1a data).

Level 1a data that have been processed to sensor units (e. g., radar backscatter cross section, brightness temperature, etc.); not all instruments have Level 1b data; level 0 data is not recoverable from level 1b data.

Derived geophysical variables (e. g., ocean wave height, soil moisture, ice concentration) at the same resolution and location as Level 1 source data.

Variables mapped on uniform spacetime grid scales, usually with some completeness and consistency (e. g., missing points interpolated, complete regions mosaicked together from multiple orbits, etc).

Model output or results from analyses of lower level data (i. e., variables that were not measured by the instruments but instead are derived from these measurements).

Level 1 data record is the most fundamental (i. e., highest reversible level) data record that has significant scientific utility, and is the foundation upon which all subsequent data sets are produced. Level 2 is the first level that is directly usable for most scientific applications; its value is much greater than the lower levels.

Level 2 data sets tend to be less voluminous than Level 1 data because they have been reduced temporally, spatially, or spectrally. Level 3 data sets are generally smaller than lower level data sets and thus can be dealt with without incurring a great deal of data handling overhead. These data tend to be generally more useful for many applications. The regular spatial and temporal organisation of Level 3 datasets makes it feasible to readily combine data from different sources.

History: The modern discipline of remote sensing arose with the development of flight. The balloonist G. Tournachon (alias Nadar) made photographs of Paris from his balloon in 1858. Messenger pigeons, kites, rockets and unmanned balloons were also used for early images. With the exception of balloons, these first, individual images were not particularly useful for map making or for scientific purposes.

Systematic aerial photography was developed for military surveillance and reconnaissance purposes beginning in World War I and reaching a climax during the Cold War with the use of modified combat aircraft such as the P-51, P-38, RB-66 and the F-4C, or specifically designed collection platforms such as the U2/TR-1, SR-71, A-5 and the OV-1 series both in overhead and standoff collection. A more recent development is that of increasingly smaller sensor pods such as those used by law enforcement and the military, in both manned and unmanned platforms. The advantage of this approach is that this requires minimal modification to a given airframe. Later imaging technologies would include infrared, conventional, doppler and synthetic aperture radar.

The development of artificial satellites in the latter half of the 20th century allowed remote sensing to progress to a global scale as of the end of the Cold War. Instrumentation aboard various Earth observing and weather satellites such as Landsat, the Nimbus and more recent missions such as RADARSAT and UARS provided global measurements of various data for civil, research, and military purposes. Space probes to other planets have also provided the opportunity to conduct remote sensing studies in extraterrestrial environments, synthetic aperture radar aboard the Magellan spacecraft provided detailed topographic maps of Venus, while instruments aboard SOHO allowed studies to be performed on the Sun and the solar wind, just to name a few examples.

Recent developments include, beginning in the 1960s and 1970s with the development of image processing of satellite imagery. Several research groups in Silicon Valley including NASA Ames Research Centre, GTE and ESL Inc. developed Fourier transform techniques leading to the first notable enhancement of imagery data.

Remote Sensing Software

Remote Sensing data is processed and analysed with computer software, known as a remote sensing application. A large number of proprietary and open source applications exist to process remote sensing data. According to an NOAA Sponsored Research by Global Marketing Insights, Inc. the most used applications among Asian academic groups involved in remote sensing are as follows:

ERDAS 36% (ERDAS IMAGINE 25% & ERMapper 11%); ESRI 30%; ITT Visual Information Solutions ENVI 17%; MapInfo 17%. Among Western Academic respondents as follows: ESRI 39%, ERDAS IMAGINE 27%, MapInfo 9%, AutoDesk 7%, ITT Visual Information Solutions ENVI 17%. Other important Remote Sensing Software packages include:

TNTmips from MicroImages, PCI Geomatica made by PCI Geomatics, the leading remote sensing software package in Canada, IDRISI from Clark Labs, Image Analyst from Intergraph, and RemoteView made by Overwatch Textron Systems. Dragon/ips is one of the oldest remote sensing packages still available, and is in some cases free. Open source remote sensing software includes GRASS GIS, ILWIS, QGIS, OSSIM, Opticks (software) and Orfeo toolbox.

Indian Remote Sensing

Indian Remote Sensing satellites (IRS) are a series of Earth Observation satellites, built, launched and maintained by Indian Space Research Organisation. The IRS series provides many remote sensing services to India.

IRS System

Following the successful demonstration flights of Bhaskara-1 and Bhaskara-2 satellites launched in 1979 and 1981, respectively, India began to develop the indigenous Indian Remote Sensing (IRS) satellite program to support the national economy in the areas of agriculture, water resources, forestry and ecology, geology, water sheds, marine fisheries and coastal management.

Towards this end, India established the National Natural Resources Management System (NNRMS) for which the Department of Space (DOS) is the nodal agency, providing operational remote sensing data services. Data from the IRS satellites is received and disseminated by several countries all over the world. With the advent of high-resolution satellites new applications in the areas of urban sprawl, infrastructure planning and other large scale applications for mapping have been initiated.

The IRS system is the largest constellation of remote sensing satellites for civilian use in operation today in the world, with 10 operational satellites. All these are placed in polar sun-synchronous orbit and provide data in a variety of spatial, spectral and temporal resolutions.

IRS Data Applications

Data from Indian Remote Sensing satellites are used for various applications of resources survey and management under the National Natural Resources Management System (NNRMS). Following is the list of those applications:

- Preharvest crop area and production estimation of major crops.
- Drought monitoring and assessment based on vegetation condition.
- Flood risk zone mapping and flood damage assessment.
- Hydro-geomorphological maps for locating underground water resources for drilling well.
- Irrigation command area status monitoring
- Snow-melt run-off estimates for planning water use in down stream projects
- Land use and land cover mapping
- Urban planning
- Forest survey
- Wetland mapping
- Environmental impact analysis
- Mineral Prospecting
- Coastal studies

Integrated Mission for Sustainable Development (initiated in 1992) for generating locale-specific prescriptions for integrated land and water resources development in 174 districts.

IRS Launch Log

The initial versions are composed of the 1 (A,B,C,D). The later versions are named based on their area of application including OceanSat, CartoSat, ResourceSat. Some of the satellites have alternate designations based on the launch number and vehicle (P series for PSLV).

S.No.	Satellite	Date of Launch	Launch Vehicle	Status
1	IRS 1A	17 March 1988	Vostok, USSR	Mission Completed
2	IRS 1B	29 August 1991	Vostok, USSR	Mission Completed
3	IRS P1 (also IE)	20 September 1993	PSLV-D1	Crashed, due to launch failure of PSLV
4	IRS P2	15 October 1994	PSLV-D2	Mission Completed
5	IRS 1C	28 December 1995	Molniya, Russia	Mission Completed
6	IRS P3	21 March 1996	PSLV-D3	Mission Completed
7	IRS 1D	29 September 1997	PSLV-C1	Mission Completed
8	IRS P4 (Oceansat-1)	27 May 1999	PSLV-C2	Mission Completed
9	Technology Experiment Satellite (TES)	22 October 2001	PSLV-C3	In Service
10	IRS P6 (Resourcesat-1)	17 October 2003	PSLV-C5	In Service
11	IRS P5 (Cartosat 1)	5 May 2005	PSLV-C6	In Service
12	Cartosat 2 (IRS P7)	10 January 2007	PSLV-C7	In Service
13	Cartosat 2A (IRS P?)	28 April 2008	PSLV-C9	In Service
14	IMS 1 (IRS P?)	28 April 2008	PSLV-C9	In Service
15	Oceansat-2	23 September 2009	PSLV-C14	In Service
16	Cartosat-2B	12 July 2010	PSLV-C15	In Service
17	Resourcesat-2	20 April 2011	PSLV-C16	In Service

Future IRS Launches

Following are the remote sensing satellites planned by ISRO to be launched next strengthening the fleet of IRS satellites and widening their applications:

- *RISAT (Radar Imaging Satellite):* A microwave remote sensing mission with Synthetic Aperture Radar (SAR) operating in C-band and having a 6 x 2 meter planar active array antenna based on trans-receiver module architecture. SAR is an all weather-imaging sensor capable of taking images in cloudy and snow covered regions and also both during day and night. RISAT weighs 1,750 kg.

- *RESEOURCESAT-3:* A follow on to Resourcesat-2, it will carry more advanced LISS-III-WS (Wide Swath) Sensor having similar swath and revisit capability as Advanced Wide Field Sensor (AWiFS), thus overcoming any spatial resolution limitation of AWiFS. Satellite would also carry Atmospheric Correction Sensor (ACS) for quantitative interpretation and geophysical parameter retrieval. It slated to be launched during 2011-12.
- *CARTOSAT-3:* A continuation of Cartosat series , it will have a resolution 30 cm and 6 km swath suitable for cadastre and infrastructure mapping and analysis. It would also enhance disaster monitoring and damage assessment. It is slated to be launched during 2011-12.
- *OCEANSAT-3:* Oceasat-3 would carry Thermal IR Sensor, 12 channel Ocean Colour Monitor, Scatterometer and Passive Microwave Radiometer. IR Sensor and Ocean Colour Monitor would be used in the analysis for operational Potential Fishing Zones. Satellite is mainly for Ocean biology and sea state applications. It is slated to the launched aboard PSLV in 2012-13.

Satellite Data Acquisition and Processing

The National Remote Sensing Centre (NRSC) at Hyderabad is the nodal agency for reception, archival, processing and dissemination of remote sensing data in the country. NRSC acquires and processes data from all Indian remote sensing satellites like CARTOSAT-1, CARTOSAT-2, RESOURCESAT-1, IRS-1D, OCEANSAT-1 and TES as well as foreign satellites like Terra, NOAA and ERS.

Remote Sensing Technology Capabilities

Applying Remote Sensing to the Kyoto Protocol: A concern with the Kyoto Protocol is the current imprecise definition of a "forest", which, in terms of ecosystem type, canopy cover, minimum area of interest etc., will have significant implications on the applicability of remote sensing technology to the treaty (Skole and Qi, paper #5). The IPCC is currently examining the implications of different forest definitions for the Protocol and is also evaluating the merits of a more quantitative approach to land cover monitoring which would focus on carbon and biomass as a basic unit of measurement. While some of these issues will be addressed at the 6th Conference of the Parties (COP-6) in The Hague (NL) in November 2000, there is a need for the remote sensing community to provide a synopsis of what Earth observations can do relative to the land cover issues as they are stated now. In this context, five areas were identified where remote sensing technology may be applied, partly or fully, toward facilitating the treaty:

- Provision of systematic observations of relevant land cover (Art. 5, Art. 10);
- Support to the establishment of a 1990 carbon stock baseline (Art. 3);
- Detection and spatial quantification of change in land cover (Art. 3, Art. 12);
- Quantification of above-ground vegetation biomass stocks and associated changes therein (Art. 3 Art 12); Mapping and monitoring of sources of anthropogenic CH4 (Art. 3, Art. 5, Art.10);

Provision of Systematic Observations of Relevant Land Cover: No later than one year prior to the start of the first commitment period, a national system for the estimation of anthropogenic emissions by sources and removals by sinks of all greenhouse gases not controlled by the Montreal Protocol." States that countries shall "Cooperate in scientific and technical research and promote the maintenance and the development of systematic observation systems and development of data archives to reduce uncertainties related to the climate system, [and] the adverse impacts of climate change...". Providing systematic, repetitive observations of large areas is potentially one of the strengths of remote sensing technology, and one where it can provide substantial support to the protocol on a long-term basis.

Remote sensing data are, however, generally not acquired in a systematic manner, except locally over specific study sites and regional scale analysis of archived data are often complicated by variations in seasonally, sensor characteristics, viewing geometry etc., which introduce biases and uncertainties in the interpretation of the results. This is typically valid for most operational sensors, both optical and microwave, thereby undermining the usefulness of the data. It is recognised that dedicated and systematic acquisition strategies, focusing on obtaining regional coverage on a repetitive basis, would significantly improve the usefulness of remote sensing data, not only in the context of the Kyoto Protocol, but also in a broader scientific framework. Although global or regional scale projects, such as the Landsat Pathfinder, TREES (Achard *et al.* 1997), GRFM/GBFM (Rosenqvist *et al.* 2000) and IGBP DIS (Belward *et al.* 1999), have existed for a long time, it is recognised that a federated approach having common goals and thematic definitions will be required to effectively support the Kyoto Protocol.

Such an effort is currently underway within the framework of the Global Observations of Forest Cover (GOFC) Pilot Project (Ahern *et al.* 1998), under the auspices of CEOS.

Passive Optical (Multi-spectral and Panchromatic) Systems: Spaceborne optical systems have been in operation since 1972 and thematic mapping applications are generally past their initial research stages. However, while results for numerous land cover mapping applications have been presented over the years, they are often site specific and focused on a particular science objective. The feasibility of identifying the thematic classes directly applicable to the Kyoto Protocol remains to be confirmed, and in some cases, further investigated.

While panchromatic systems are of limited use for thematic mapping of vegetation, multispectral systems, in particular sensors which include mid-infrared bands such as Landsat TM, ETM+ and SPOT HRVIR, are well suited for this purpose. High resolution data will be required for the delineation of fragmented forest lands and smaller patches of forest.

Coarser resolution sensors such as NOAA AVHRR and SPOT VEGETATION are frequently used in combination with high resolution sensors for continental and global scale mapping Currently available optical systems are generally capable of acquiring data at local, regional, and global scales and in a timely and regular manner. Cloud cover, smoke and haze, however put limitations on data availability, particularly in the tropical zone. The problem can be somewhat overcome with the coarse resolution sensors which have a higher temporal repeat-cycle, thereby enabling the creation of weekly or monthly mosaics compiled from cloud free pixels.

Active Microwave Systems (SAR): The current suite of orbiting SAR systems all operate with a single-band and a single polarization, which to a large extent limit their usefulness for thematic mapping of vegetation Classification accuracy with SAR, however, increases notably with the inclusion of additional bands and polarizations, and with multi-temporal data acquisitions. Very good results can also be achieved when optical and microwave data are combined. While multi-band/polarimetric and interferometric radar systems are not yet available in a space-borne mode, they are available on aircraft platforms and could be used for local to regional scale applications.

A major limitation of radar systems, with respect to vegetation mapping, is their sensitivity to surface topography which limits their application to flat or gently undulating terrain. Radar data are also subject to speckle, which on one hand enables techniques such as radar interferometry, but reduces the effective ground resolution (typically by a factor 3-4, thus providing an effective resolution in the order of 50-100 metres for current spaceborne systems).

For ultra-wide band radar systems, however, the speckle problem is absent. The advantage of radar systems is their all-weather capability, which assures image acquisitions independent of cloud cover and daylight, thereby enabling timely and reliable acquisitions at local, regional and global scales. It should however be noted that radar systems are not "weather independent" as hydrologic conditions on the ground such as; wet, flooded, or snow covered soil affect the radar signal.

Active Optical Systems (LIDAR): LIDAR systems are only just recently being explored for vegetation mapping. At the time of this writing, NASA's Vegetation Canopy LIDAR (VCL) is the only LIDAR system planned for orbit in the near future. VCL is an active infrared laser altimeter which will make soundings of the vegetation canopy, providing unprecedented information on the structure of the Earth's forests and land surfaces by directly observing vegetation canopy height, forest vertical and spatial distribution, and ground topography at high resolution.

VCL is however not an imaging instrument. It will collect data in a series of samples, along the flight path. As such the production of thematic products from VCL is, as of yet, unproven. However, using VCL data in combination with other spatially extensive data, such as optical/multispectral or SAR, holds a significant potential.

Support to the Establishment of a 1990 Carbon Stock Baseline: According to Art. 3:4 of the Kyoto Protocol, each Annex I country shall "provide data to establish its level of carbon stocks in 1990 and to enable an estimate to be made of its changes in carbon stocks in subsequent years".

However, Art. 3:5 of the Protocol also states that Annex I countries "undergoing the process of transition to a market economy" may, under certain circumstances, "use a historical base year or period other than 1990 for the implementation of its commitments" under Art. 3. Hence, baselines formulated after 1990 may, for certain countries, be considered. Nevertheless, as it can be expected that the year 1990 will be the dominant base year, the selection of potential sensors to be used to support the establishment of this base line will to the largest extent be limited to those in operation during that year.

Among the high resolution optical sensors, only Landsat TM and SPOT HRV were in operation in 1990. The use of high resolution data for compiling a regional-global 1990 land cover map to support the establishment of the carbon stock baseline is possible - albeit expensive. It is feasible at a national level, especially for smaller countries or regions. Archives of Landsat TM and MSS, and SPOT HRV exist and could be used for this purpose.

The use of coarse resolution data is also feasible, although spatial resolution issues for many areas would limit its utility. A Global Land Cover map from 1992 has been generated from NOAA AVHRR data within IGBP DIS and archives of NOAA AVHRR data exist for the required time period. In order to be useful, however, the land cover classes used need to be redefined and adapted to classes relevant to the Kyoto Protocol.

No orbital active microwave systems were in operation in 1990 and the use of SAR data to support a 1990 baseline would thus in general not be feasible. SAR data could however possibly be useful for quantification of component biomass (leaves, branches, stems) of the extensive areas of woodlands that occur throughout Australia, Africa and South America, for which the establishment of a 1990 baseline could be supported.

For non-1990 baseline countries in the tropical and boreal zones of the Earth, continental scale (100 m resolution) JERS-1 L-band SAR mosaics from 1995-96, generated within the GRFM/GBFM projects (Rosenqvist *et al.* 2000), can be used to support the establishment of a mid-1990's carbon baseline.

In the first commitment period (2008-2012) this application primarily concerns the detection and spatial quantification of afforestation, reforestation and deforestation (ARD) activities, and changes resulting from fire. While the accounting of land cover change is initially limited to that caused by human activity, disturbances and changes due to natural causes also need to be identified. In the second and subsequent commitment periods, accounting will also include other types of land use change.

The treaty states that "The net changes in greenhouse gas emissions by sources and removals by sinks resulting from direct human-induced land-use change and forestry activities, limited to afforestation, reforestation and deforestation since 1990, measured as verifiable changes in carbon stocks in each commitment period, This, in turn, would require verifiable measurements of ARD, should LULUCF projects (Land Use, Land Use Change and Forestry) be accepted under the terms of the clean development mechanism.

While the articles above concern measurements of carbon stocks and changes therein, a first important step is the identification and quantification of the areas subject to ARD. In combination with up-to-date *in situ* data and relevant allometric models, changes in biomass(carbon) stocks may be estimated.

In order to detect ARD activities, image acquisitions on a repetitive basis will be required, preferably annually and performed during a specific season, in order to minimise the effects of seasonal artefacts in the data. A spatial resolution better than the minimum area of interest - still to be defined - will be required for this task.

Optical systems are sensitive to parameters related to the structure and closure of the vegetation canopy, (e.g. canopy projected cover (CPC) and leaf area index (LAI)) which are affected during ARD activities and fire. Detection and spatial quantification of deforestation (D) activities, which bring about the removal of the forest canopy, is the most straight-forward part of the three ARD components, and both panchromatic and multi-spectral remote sensing data are deemed useful for this task. High resolution systems will be required to detect partial deforestation activities, such as selective logging and thinning. Reforestation (R) is more difficult to detect, as it represents a gradual change from non-forest to forest, spanning several years. Afforestation (A) events, which can be expected to take place in relatively small patches outside the "forest" areas will be most difficult to detect of the ARD components. Multi-spectral systems are however sensitive to growth parameters such as APAR (absorbed photosynthetically active radiation), which peaks during the regeneration stages, thus indicating the location of potential R and A areas after the trees are large enough.

High resolution multi-spectral systems will be required for both R and A, but repetitive (annual) measurements will be essential. Persistent cloud coverage in some areas constitutes a major obstacle. Nevertheless, the simple identification that ARD activities have taken place over time can be achieved (Justice *et al.* 1996) and is a valid contribution in the context of the protocol Active fire events can be detected in an operational manner both at global and regional scale by coarse resolution optical sensors, which provide daily coverage. Spatial quantification of the burnt areas can thereafter be assessed with the use of high resolution sensors.

The World Fire Web network provides near-real time information on global fire activities using NOAA AVHRR data (Pinnock and Grégoire, 2000).

Microwave sensors are particularly sensitive to detecting changes between images acquired at different times, even in areas of topographic terrain, provided that the viewing geometry is kept the same. Instruments operating with long wavelengths (L-band or longer) are more suitable for forest related monitoring than short wavelength sensors (C-band or shorter) as the L-band signals interact with the forest at branch and trunk level, while the main interaction at C-band occurs with the canopy. Rough soil and herbaceous vegetation may, in the latter case, be confused with forest.

Multi-baseline interferometry, however, even at C-band may be used to separate the contributions of ground and canopy for many forest types (Treuhaft and Siqueira 2000). Tomographic multi-baseline approaches (Reigber and Moreira, in press) will also play an important role in separating ground and volume contributions, and therefore will apply to detecting change in forest area. Since SAR image acquisitions are independent of cloud cover, it is possible to accurately plan the timing of the data takes, thus optimising the conditions to detect change in the land cover.

While it is possible to use short wavelength band SAR systems for the detection and spatial quantification of deforestation (D) events, the use of single polarization C-band data has proven to be problematic as forest and non-forest areas cannot always be differentiated. Still, interferometric C-band tandem data, in particular the phase coherence, may under certain circumstances constitute a valuable source of land cover type information. If limited to single polarization data sets, the detection and quantification of deforested (D), reforested (R) and afforested (A) areas is best addressed using longer wavelength band SAR systems, which are more sensitive to the range of biomass associated with forests.

Active fires are not possible to detect with microwave systems as the smoke plumes are invisible to the radar. Burn scars however, may be detected in cases where the fire has caused substantial change to the structure of the forest (Antikidis *et al.* 1997) and can be detected from SAR for several years after the burn.

Polarimetric SAR and polarimetric interferometric systems (Cloude and Papathanassiou 1998) will improve the capabilities for ARD monitoring (Kellndorfer et al. 1998) and at least three such systems are currently planned for the near future: ALOS (L-band), Envisat (C-band) and Radarsat-2 (C-band). The LightSAR (L-band) mission has been halted, but NASA is currently studying alternative mission concepts.

As long as imaging LIDAR systems are unavailable, detection of ARD activities and burn scars will belimited to those areas actually sampled by the VCL. As such, the feasibility of using LIDAR to address ARD events for extensive areas is, as of yet, unproven. However, LIDAR should have the ability to repeatedly characterise structural attributes at specific locations or collect sample sets in known ARD areas which could prove useful (Blair *et al.* 1999).

Direct measurements of total above ground forest biomass stocks or changes in such is not feasible with (passive) optical systems. However, indirect estimations of biomass change is possible to a limited extent using vegetation indices based on photosynthetically active radiation (PAR) or through indirect relationships with, for example, mid infrared reflectance data.

PAR measurements have been routinely made using multi-spectral sensors and may be combined with environmental data and forest growth models to predict NPP (net primary production) which is presented in terms of units of carbon [e.g. kgC /ha/yr]. Future perspectives in this domain through the exploitation of multiangular measurements, e.g. from the MISR instrument, also offer a significantf potential (Martonchik *et al.* 1998).

The application of radar systems to measure and detect changes in above-ground biomass stocks is an active area of research and development. There was general agreement among the workshop participants, that currently available radar satellite systems (ERS-2 and Radarsat-1), which operate with single channel C-band, are not well suited for biomass estimation, since the signals saturate at low biomass levels. Although C-band sensitivity to biomass up to almost 100 t/ha in certain circumstances may be achieved by fixed-baseline and dual-pass interferometric techniques, the accuracy is largely dependent on factors such as the baseline distance and surface conditions, and more R&D is required before the technique may become operational with single band data.

Polarimetric interferometry at L- and C-band, however, is a field of research which holds a certain potential for biomass monitoring L-band SAR, with a biomass saturation level of 60-100 tons/ha, may be useful for coarse biomass estimates in regeneration areas (A and R components). There are currently no L-band SAR systems in orbit (JERS-1 failed 1998), but a polarimetric Lband system is planned for the ALOS satellite (due for launch in 2002) and could be well suited to address biomass issues in the context of the Kyoto Protocol. Interferometric coherence by Lband SAR is yet to be investigated.

While the biomass levels approachable by L-band SAR are still way below those of mature forests, which vary between 100 - 600 t/ha, longer wavelengths, together with polarimetric and/or interferometric techniques, can be used to push the biomass saturation levels forward and to improve accuracy (Dobson *et al.* 1992, Imhoff 1995). Aircraft based radar sensors having full multi-band, polarimetric, and interferometric capabilities currently exist and have proven capable of detecting biomass in a wide range of forests up to 200 tons/ha dry above ground biomass (Ranson *et al.* 1997), primarily due to the long wavelength Pband channel.

No space-borne P-band SAR system has been launched up-to-date, mainly due to unresolved ionospheric effects associated with low frequency radars. These effects, which are functions of the total electron content (TEC) in the ionosphere, result in deformation and polarimetric rotation of the signal. It has however recently been shown that polarimetric (Faraday) rotation may be corrected for by using a fully polarimetric system (Freeman *et al.* 1998), while other ionospheric artefacts may be reduced to acceptable levels by accurate timing of the data acquisition (early dawn) when TEC is at minimum (Siqueira *et al.*, paper #10).

Physical constraints in instrument design, relating to minimum antenna dimensions, can be bypassed at the expense of a lower (50-100 m) ground resolution (Freeman et al. 1999). Lower frequency SAR systems (VHF band) are probably the most useful for direct measurement and mapping of biomass. In this frequency band, the forest trunks act as Rayleigh scatterers and accordingly has a linear relationship between radar amplitude and stem volume. The VHF-band sensitivity to surface slopes is much less than at higher frequencies, and can to a large extent be corrected for (Smith and Ulander 2000). Recent studies have shown that these frequencies are capable of accurately measuring biomass above 100 tons/ha (dry above ground biomass). Results from the aircraft based CARABAS (20-90 MHz; Ulander et al. 1998), deployed in temperate and boreal forests in Europe, and the BioSAR system (80-120 MHz, Imhoff et al. 2000), deployed in the neo-tropics, have shown that biomass measures can be accurately derived (within ± 10% of field measures) for forests between 100 and 500 tons/ha (actual saturation levels for the BioSAR and CARABAS systems have yet to be determined).

These systems show great promise for local to regional applications using aircraft. However, the deployment of space-based VHF/UHF sensors may not be technically feasible due to ionospheric interference with the signal and has yet to be explored (Ishimaru *et al.* 1999). Alternative platforms may also be considered, including long endurance stratospheric airships (UAV) which are not subject to ionospheric distortions. Such platforms will in the near future have an endurance of 6 months to 5 years, through use of combinations of solar cell and fuel cell technologies.

The possibility of using UHF and VHF radar for routine forest biomass measurement is very real. Technological advancements are eliminating many of the obstacles that previously limited the development of UHF and VHF systems and orbital systems may be possible in the near future. In order to take advantage of the potential of these systems, , the scientific community needs to make the appropriate frequency allocation requirements known to the International Telecommunications Union (ITU) and the World Radio Conference (WCR) so that some part of the spectrum can be set aside for remote sensing purposes. Combined with allometric models (models linking biomass to measurable parameters such as tree height) the data collected by VCL should be capable of helping it make accurate measures of above ground biomass based on vegetation structure and canopy height measures. Combined with spatially extensive data, such as optical or SAR imagery, interpolation of biomass estimates between VCL sample points could be used to provide local, specific site, biomass estimates. As mentioned previously, it remains to be seen how such data will be applied over large areas. A limitation of LIDAR is that species/genera cannot be discriminated and yet wood density - and hence biomass - may vary considerably between different species of similar height and similar age. Environmental factors also affect biomass/height relationships.

Mapping and Monitoring of Certain Sources of Anthropogenic $CH_{4:}$ The Kyoto Protocol states that " The Parties included in Annex I shall, individually or jointly, ensure that their aggregate anthropogenic carbon dioxide equivalent emissions of the greenhouse gases listed in Annex A do not exceed their assigned amounts..." , "with a view to reducing their overall emissions of such gases by at least 5 per cent below 1990 levels in the commitment period 2008 to 2012". Hence, the Kyoto Protocol relates to all six greenhouse gases listed in Annex A, including CH_4, which is the second of the two greenhouse gases, apart from CO_2, considered relevant in the context of this report.

Although it may well be included above, mapping and monitoring of certain sources of anthropogenic CH_4 is here listed separately, as it is not generally related to forestry orto forest change. Apart from livestock management - which is not considered feasible to monitor (Principles V and VIII); by remote sensing CH_4 is also emitted as a result of anaerobic conditions in open water bodies following extended inundation. Typical anthropogenic sources of CH_4 include irrigated rice paddies, aquaculture (e.g. fish- and shrimp cultivation) and hydroelectric reservoirs. Quantification of global biomass burning patterns and burnt area will also help to close uncertainties in CH4 emissions from vegetation fires. Detection and spatial quantification of open water bodies such as aquaculture and hydroelectric reservoirs can be performed by single date high resolution optical sensors. Rice paddies may also be identified in a single date, although repetitive measurements during the growth season are recommended in order to be able to separate it from other agricultural crops and to monitor the water regime - the key factor triggering the CH_4 emissions. However, cloud cover will, in the latter case, constitute an obstacle to obtaining the relevant multi-temporal data.

SAR data can be used to map the three CH4 sources referred to above. SAR data is particularly suitable for multi-temporal monitoring of irrigated rice, as regular acquisitions can be performed irrespectable of the cloud conditions. Both C-band and L-band SAR have been used to map rice growth and it is now deemed possible to perform this in an operational way, using current sensors (ERS-2 and Radarsat-1) as well as sensors planned in the near future (ALOS PALSAR, ENVISAT ASAR, Radarsat-2).

The feasibility of using LIDAR to address this issue is not currently known.

Chapter 3

Geographic Information System

A geographic information system (GIS), geographical information system, or geospatial information system is any system that captures, stores, analyses, manages, and presents data that are linked to location(s). In the simplest terms, GIS is the merging of cartography, statistical analysis, and database technology. GIS may be used in archaeology, geography, cartography, remote sensing, land surveying, public utility management, natural resource management, precision agriculture, photogrammetry, urban planning, emergency management, navigation, aerial video, and localized search engines.

As GIS can be thought of as a system, it digitally creates and "manipulates" spatial areas that may be jurisdictional, purpose or application oriented for which a specific GIS is developed. Hence, a GIS developed for an application, jurisdiction, enterprise, or purpose may not be necessarily interoperable or compatible with a GIS that has been developed for some other application, jurisdiction, enterprise, or purpose. What goes beyond a GIS is a spatial data infrastructure (SDI), a concept that has no such restrictive boundaries.

Therefore, in a general sense, the term describes any information system that integrates, stores, edits, analyses, shares, and displays geographic information for informing decision making. GIS applications are tools that allow users to create interactive queries (user-created searches), analyse spatial information, edit data, maps, and present the results of all these operations. Geographic information science is the science underlying the geographic concepts, applications and systems. GIS can be studied in degree and certificate programs at many universities.

Applications

GIS technology can be used for: earth surface based scientific investigations; resource management, reference, and projections of a geospatial nature—both manmade and natural; asset management and location planning; archaeology; environmental impact study; infrastructure assessment and development; urban planning; cartography, for a thematic and/or time based purpose; criminology; GIS data development geographic history; marketing; logistics; population and demographic studies; prospectivity mapping; location attributes applied statistical analysis; warfare assessments; and other purposes. Examples of use are: GIS may allow emergency planners to easily calculate emergency response times and the movement of response resources (for logistics) in the case of a natural disaster; GIS might be used to find wetlands that need protection strategies regarding pollution; or GIS can be used by a company to site a new business location to take advantage of GIS data identified trends to respond to a previously under-served market. Most city and transportation systems planning offices have GIS sections.

History of Development

In 1854, John Snow depicted a cholera outbreak in London using points to represent the locations of some individual cases, possibly the earliest use of the geographic method. This is bad. His study of the distribution of cholera led to the source of the disease, a contaminated water pump (the Broad Street Pump, whose handle he had disconnected, thus terminating the outbreak) within the heart of the cholera outbreak.

While the basic elements of topography and theme existed previously in cartography, the John Snow map was unique, using cartographic methods not only to depict but also to analyse clusters of geographically dependent phenomena for the first time.

The early 20th century saw the development of photolithography, by which maps were separated into layers. Computer hardware development spurred by nuclear weapon research led to general-purpose computer "mapping" applications by the early 1960s.

The year 1962 saw the development of the world's first true operational GIS in Ottawa, Ontario, Canada by the federal Department of Forestry and Rural Development. Developed by Dr. Roger Tomlinson, it was called the "Canada Geographic Information System" (CGIS) and was used to store, analyse, and manipulate data collected for the Canada Land Inventory (CLI) – an effort to determine the land capability for rural Canada by mapping information about soils, agriculture, recreation, wildlife, waterfowl, forestry, and land use at a scale of 1:50,000. A rating classification factor was also added to permit analysis.

CGIS was the world's first such system and an improvement over "mapping" applications as it provided capabilities for overlay, measurement, and digitizing/scanning. It supported a national coordinate system that spanned the continent, coded lines as "arcs" having a true embedded topology, and it stored the attribute and locational information in separate files. As a result of this, Tomlinson has become known as the "father of GIS," particularly for his use of overlays in promoting the spatial analysis of convergent geographic data. CGIS lasted into the 1990s and built a large digital land resource database in Canada. It was developed as a mainframe based system in support of federal and provincial resource planning and management. Its strength was continent-wide analysis of complex datasets. The CGIS was never available in a commercial form.

In 1964, Howard T Fisher formed the Laboratory for Computer Graphics and Spatial Analysis at the Harvard Graduate School of Design (LCGSA 1965-1991), where a number of important theoretical concepts in spatial data handling were developed, and which by the 1970s had distributed seminal software code and systems, such as 'SYMAP', 'GRID', and 'ODYSSEY' — which served as literal and inspirational sources for subsequent commercial development—to universities, research centres, and corporations worldwide.

By the early 1980s, M&S Computing (later Intergraph), Environmental Systems Research Institute (ESRI), CARIS (Computer Aided Resource Information System) and ERDAS emerged as commercial vendors of GIS software, successfully incorporating many of the CGIS features, combining the first generation approach to separation of spatial and attribute information with a second generation approach to organizing attribute data into database structures. In parallel, the development of two public domain systems began in the late 1970s and early 1980s. MOSS, the Map Overlay and Statistical System project started in 1977 in Fort Collins, Colorado under the auspices of the Western Energy and Land Use Team (WELUT) and the U.S. Fish and Wildlife Service. GRASS GIS was begun in 1982 by the U.S. Army Corps of Engineering Research Laboratory (USA-CERL) in Champaign, Illinois, a branch of the U.S. Army Corps of Engineers to meet the need of the U.S. military for software for land management and environmental planning. The later 1980s and 1990s industry growth were spurred on by the growing use of GIS on Unix workstations and the personal computer. By the end of the 20th century, the rapid growth in various systems had been consolidated and standardized on relatively few platforms, and users were beginning to export the concept of viewing GIS data over the Internet, requiring data format and transfer standards. More recently, a growing number of free, open source GIS packages run on a range of operating systems and can be customized to perform specific tasks. Increasingly geospatial data and mapping applications are being made available via the world wide web.

Several authoritative books on the history of GIS have been published.

GIS Techniques and Technology

Modern GIS technologies use digital information, for which various digitized data creation methods are used. The most common method of data creation is digitization, where a hard copy map or survey plan is transferred into a digital medium through the use of a computer-aided design (CAD) program, and geo-referencing capabilities. With the wide availability of ortho-rectified imagery (both from satellite and aerial sources), heads-up digitizing is becoming the main avenue through which geographic data is extracted. Heads-up digitizing involves the tracing of geographic data directly on top of the aerial imagery instead of by the traditional method of tracing the geographic form on a separate digitizing tablet (heads-down digitizing).

Relating Information from Different Sources

GIS uses spatio-temporal (space-time) location as the key index variable for all other information. Just as a relational database containing text or numbers can relate many different tables using common key index variables, GIS can relate otherwise unrelated information by using location as the key index variable. The key is the location and/or extent in space-time.

Any variable that can be located spatially, and increasingly also temporally, can be referenced using a GIS. Locations or extents in Earth space-time may be recorded as dates/times of occurrence, and x, y, and z coordinates representing, longitude, latitude, and elevation, respectively. These GIS coordinates may represent other quantified systems of temporo-spatial reference (for example, film frame number, stream gage station, highway mile marker, surveyor benchmark, building address, street intersection, entrance gate, water depth sounding, POS or CAD drawing origin/units). Units applied to recorded temporal-spatial data can vary widely, but all Earth-based spatial-temporal location and extent references should, ideally, be relatable to one another and ultimately to a "real" physical location or extent in space-time.

Related by accurate spatial information, an incredible variety of real-world and projected past or future data can be analysed, interpreted and represented to facilitate education and decision making. This key characteristic of GIS has begun to open new avenues of scientific inquiry into Behaviours and patterns of previously considered unrelated real-world information.

GIS Uncertainties

GIS accuracy depends upon source data, and how it is encoded to be data referenced. Land Surveyors have been able to provide a high level of positional accuracy utilizing the GPS derived positions. [Retrieved from Federal Geographic Data Committee] the high-resolution digital terrain and aerial imagery, [Retrieved NJGIN] the powerful computers, Web technology, are changing the quality, utility, and expectations of GIS to serve society on a grand scale, but nevertheless there are other source data that has an impact on the overall GIS accuracy like: paper maps that are not found to be very suitable to achieve the desired accuracy since the aging of maps affects their dimensional stability.

In developing a Digital Topographic Data Base for a GIS, topographical maps are the main source of data. Aerial photography and satellite images are extra sources for collecting data and identifying attributes which can be mapped in layers over a location facsimile of scale. The scale of a map and geographical rendering area representation type are a very important aspects since the information content depends mainly on the scale set and resulting locatability of the map's representations. In order to digitize a map, the map has to be checked within theoretical dimensions, then scanned into a raster format, and resulting raster data has to be given a theoretical dimension by a rubber sheeting/warping technology process.

Uncertainty is a significant problem in designing a GIS because spatial data tend to be used for purposes for which they were never intended. Some maps were made many decades ago, where at that time the computer industry was not even in its perspective establishments. This has led to historical reference maps without common norms. Map accuracy is a relative issue of minor importance in cartography.

All maps are established for communication ends. Maps use a historically constrained technology of pen and paper to communicate a view of the world to their users. Cartographers feel little need to communicate information based on accuracy, for when the same map is digitized and input into a GIS, the mode of use often changes. The new uses extend well beyond a determined domain for which the original map was intended and designed.

A quantitative analysis of maps brings accuracy issues into focus. The electronic and other equipment used to make measurements for GIS is far more precise than the machines of conventional map analysis. [Retrieved USGS]. The truth is that all geographical data are inherently inaccurate, and these inaccuracies will propagate through GIS operations in ways that are difficult to predict, yet have goals of conveyance in mind for original design. Accuracy Standards for 1:24000 Scales Map: 1:24,000 ± 40.00 feet

This means that when we see a point or attribute on a map, its "probable" location is within a +/- 40 foot area of its rendered reference, according to area representations and scale.

A GIS can also convert existing digital information, which may not yet be in map form, into forms it can recognize, employ for its data analysis processes, and use in forming mapping output. For example, digital satellite images generated through remote sensing can be analysed to produce a map-like layer of digital information about vegetative covers on land locations. Another fairly recently developed resource for naming GIS location objects is the Getty Thesaurus of Geographic Names (GTGN), which is a structured vocabulary containing about 1,000,000 names and other information about places.

Likewise, researched census or hydrological tabular data can be displayed in map-like form, serving as layers of thematic information for forming a GIS map.

Data Representation

GIS data represents real objects (such as roads, land use, elevation, trees, waterways, etc.) with digital data determining the mix. Real objects can be divided into two abstractions: discrete objects (e.g., a house) and continuous fields (such as rainfall amount, or elevations). Traditionally, there are two broad methods used to store data in a GIS for both kinds of abstractions mapping references: raster images and vector. Points, lines, and polygons are the stuff of mapped location attribute references. A new hybrid method of storing data is that of identifying point clouds, which combine three-dimensional points with RGB information at each point, returning a "3D colour image". GIS Thematic maps then are becoming more and more realistically visually descriptive of what they set out to show or determine.

Raster

A raster data type is, in essence, any type of digital image represented by reducible and enlargeable grids. Anyone who is familiar with digital photography will recognize the Raster graphics pixel as the smallest individual grid unit building block of an image, usually not readily identified as an artifact shape until an image is produced on a very large scale.

A combination of the pixels making up an image colour formation scheme will compose details of an image, as is distinct from the commonly used points, lines, and polygon area location symbols of scalable vector graphics as the basis of the vector model of area attribute rendering. While a digital image is concerned with its output blending together its grid based details as an identifiable representation of reality, in a photograph or art image transferred into a computer, the raster data type will reflect a digitized abstraction of reality dealt with by grid populating tones or objects, quantities, cojoined or open boundaries, and map relief schemas.

Aerial photos are one commonly used form of raster data, with one primary purpose in mind: to display a detailed image on a map area, or for the purposes of rendering its identifiable objects by digitization. Additional raster data sets used by a GIS will contain information regarding elevation, a digital elevation model, or reflectance of a particular wavelength of light, Landsat, or other electromagnetic spectrum indicators. Raster data type consists of rows and columns of cells, with each cell storing a single value. Raster data can be images (raster images) with each pixel (or cell) containing a colour value. Additional values recorded for each cell may be a discrete value, such as land use, a continuous value, such as temperature, or a null value if no data is available. While a raster cell stores a single value, it can be extended by using raster bands to represent RGB (red, green, blue) colours, colormaps (a mapping between a thematic code and RGB value), or an extended attribute table with one row for each unique cell value. The resolution of the raster data set is its cell width in ground units.

Raster data is stored in various formats; from a standard file-based structure of TIF, JPEG, etc. to binary large object (BLOB) data stored directly in a relational database management system (RDBMS) similar to other vector-based feature classes. Database storage, when properly indexed, typically allows for quicker retrieval of the raster data but can require storage of millions of significantly sized records.

Vector

In a GIS, geographical features are often expressed as vectors, by considering those features as geometrical shapes. Different geographical features are expressed by different types of geometry:

Points

Zero-dimensional points are used for geographical features that can best be expressed by a single point reference — in other words, by simple location. Examples include wells, peaks, features of interest, and trailheads. Points convey the least amount of information of these file types. Points can also be used to represent areas when displayed at a small scale. For example, cities on a map of the world might be represented by points rather than polygons. No measurements are possible with point features.

Lines or Polylines

One-dimensional lines or polylines are used for linear features such as rivers, roads, railroads, trails, and topographic lines. Again, as with point features, linear features displayed at a small scale will be represented as linear features rather than as a polygon. Line features can measure distance.

Polygons

Two-dimensional polygons are used for geographical features that cover a particular area of the earth's surface. Such features may include lakes, park boundaries, buildings, city boundaries, or land uses. Polygons convey the most amount of information of the file types. Polygon features can measure perimeter and area.

Each of these geometries is linked to a row in a database that describes their attributes. For example, a database that describes lakes may contain a lake's depth, water quality, pollution level. This information can be used to make a map to describe a particular attribute of the dataset. For example, lakes could be coloured depending on level of pollution. Different geometries can also be compared. For example, the GIS could be used to identify all wells (point geometry) that are within one kilometre of a lake (polygon geometry) that has a high level of pollution.

Vector features can be made to respect spatial integrity through the application of topology rules such as 'polygons must not overlap'. Vector data can also be used to represent continuously varying phenomena. Contour lines and triangulated irregular networks (TIN) are used to represent elevation or other continuously changing values. TINs record values at point locations, which are connected by lines to form an irregular mesh of triangles. The face of the triangles represent the terrain surface.

Advantages and Disadvantages

There are some important advantages and disadvantages to using a raster or vector data model to represent reality:

- Raster datasets record a value for all points in the area covered which may require more storage space than representing data in a vector format that can store data only where needed.
- Raster data allows easy implementation of overlay operations, which are more difficult with vector data.
- Vector data can be displayed as vector graphics used on traditional maps, whereas raster data will appear as an image that may have a blocky appearance for object boundaries. (depending on the resolution of the raster file)
- Vector data can be easier to register, scale, and re-project, which can simplify combining vector layers from different sources.
- Vector data is more compatible with relational database environments, where they can be part of a relational table as a normal column and processed using a multitude of operators.

- Vector file sizes are usually smaller than raster data, which can be 10 to 100 times larger than vector data (depending on resolution).
- Vector data is simpler to update and maintain, whereas a raster image will have to be completely reproduced. (Example: a new road is added).
- Vector data allows much more analysis capability, especially for "networks" such as roads, power, rail, telecommunications, etc. (Examples: Best route, largest port, airfields connected to two-lane highways). Raster data will not have all the characteristics of the features it displays.

Non-spatial Data

Additional non-spatial data can also be stored along with the spatial data represented by the coordinates of a vector geometry or the position of a raster cell. In vector data, the additional data contains attributes of the feature. For example, a forest inventory polygon may also have an identifier value and information about tree species. In raster data the cell value can store attribute information, but it can also be used as an identifier that can relate to records in another table.

Software is currently being developed to support spatial and non-spatial decision-making, with the solutions to spatial problems being integrated with solutions to non-spatial problems. The end result with these Flexible Spatial Decision-Making Support Systems (FSDSS) is expected to be that non-experts will be able to use GIS, along with spatial criteria, and simply integrate their non-spatial criteria to view solutions to multi-criteria problems. This system is intended to assist decision-making.

Data Capture

Data capture—entering information into the system—consumes much of the time of GIS practitioners. There are a variety of methods used to enter data into a GIS where it is stored in a digital format.

Existing data printed on paper or PET film maps can be digitized or scanned to produce digital data. A digitizer produces vector data as an operator traces points, lines, and polygon boundaries from a map. Scanning a map results in raster data that could be further processed to produce vector data.

Survey data can be directly entered into a GIS from digital data collection systems on survey instruments using a technique called

Coordinate Geometry (COGO). Positions from a Global Navigation Satellite System (GNSS) like Global Positioning System (GPS), another survey tool, can also be directly entered into a GIS. Current trend is data collection and field mapping carried out directly with field computers (position from GPS and/or laser rangefinder). New technologies allow to create maps as well as analysis directly in the field, projects are more efficient and mapping is more accurate.

Remotely sensed data also plays an important role in data collection and consist of sensors attached to a platform. Sensors include cameras, digital scanners and LIDAR, while platforms usually consist of aircraft and satellites.

The majority of digital data currently comes from photo interpretation of aerial photographs. Soft copy workstations are used to digitize features directly from stereo pairs of digital photographs. These systems allow data to be captured in two and three dimensions, with elevations measured directly from a stereo pair using principles of photogrammetry. Currently, analog aerial photos are scanned before being entered into a soft copy system, but as high quality digital cameras become cheaper this step will be skipped.

Satellite remote sensing provides another important source of spatial data. Here satellites use different sensor packages to passively measure the reflectance from parts of the electromagnetic spectrum or radio waves that were sent out from an active sensor such as radar. Remote sensing collects raster data that can be further processed using different bands to identify objects and classes of interest, such as land cover.

When data is captured, the user should consider if the data should be captured with either a relative accuracy or absolute accuracy, since this could not only influence how information will be interpreted but also the cost of data capture.

In addition to collecting and entering spatial data, attribute data is also entered into a GIS. For vector data, this includes additional information about the objects represented in the system.

After entering data into a GIS, the data usually requires editing, to remove errors, or further processing. For vector data it must be made "topologically correct" before it can be used for some advanced analysis. For example, in a road network, lines must connect with nodes at an intersection. Errors such as undershoots and overshoots must also be removed. For scanned maps, blemishes on the source map may need to be removed from the resulting raster. For example, a fleck of dirt might connect two lines that should not be connected.

Raster-to-vector Translation

Data restructuring can be performed by a GIS to convert data into different formats. For example, a GIS may be used to convert a satellite image map to a vector structure by generating lines around all cells with the same classification, while determining the cell spatial relationships, such as adjacency or inclusion.

More advanced data processing can occur with image processing, a technique developed in the late 1960s by NASA and the private sector to provide contrast enhancement, false colour rendering and a variety of other techniques including use of two dimensional Fourier transforms.

Since digital data is collected and stored in various ways, the two data sources may not be entirely compatible. So a GIS must be able to convert geographic data from one structure to another.

Projections, Coordinate Systems and Registration

A property ownership map and a soils map might show data at different scales. Map information in a GIS must be manipulated so that it registers, or fits, with information gathered from other maps. Before the digital data can be analysed, they may have to undergo other manipulations—projection and coordinate conversions, for example—that integrate them into a GIS.

The earth can be represented by various models, each of which may provide a different set of coordinates (e.g., latitude, longitude, elevation) for any given point on the Earth's surface. The simplest model is to assume the earth is a perfect sphere. As more measurements of the earth have accumulated, the models of the earth have become more sophisticated and more accurate. In fact, there are models that apply to different areas of the earth to provide increased accuracy (e.g., North American Datum, 1927 - NAD27 - works well in North America, but not in Europe).

Projection is a fundamental component of map making. A projection is a mathematical means of transferring information from a model of the Earth, which represents a three-dimensional curved surface, to a two-dimensional medium—paper or a computer screen. Different projections are used for different types of maps because each projection particularly suits specific uses. For example, a projection that accurately represents the shapes of the continents will distort their relative sizes.

Since much of the information in a GIS comes from existing maps, a GIS uses the processing power of the computer to transform digital information, gathered from sources with different projections and/or different coordinate systems, to a common projection and coordinate system. For images, this process is called rectification.

Spatial Analysis with GIS

Given the vast range of spatial analysis techniques that have been developed over the past half century, any summary or review can only cover the subject to a limited depth. This is a rapidly changing field, and GIS packages are increasingly including analytical tools as standard built-in facilities or as optional toolsets, add-ins or 'analysts'. In many instances such facilities are provided by the original software suppliers (commercial vendors or collaborative non commercial development teams), whilst in other cases facilities have been developed and are provided by third parties. Furthermore, many products offer software development kits (SDKs), programming languages and language support, scripting facilities and/or special interfaces for developing one's own analytical tools or variants. The website Geospatial Analysis and associated book/ebook attempt to provide a reasonably comprehensive guide to the subject. The impact of these myriad paths to perform spatial analysis create a new dimension to business intelligence termed "spatial intelligence" which, when delivered via intranet, democratizes access to operational sorts not usually privy to this type of information.

Data Modelling

It is difficult to relate wetlands maps to rainfall amounts recorded at different points such as airports, television stations, and high schools. A GIS, however, can be used to depict two- and three-dimensional characteristics of the Earth's surface, subsurface, and atmosphere from information points. For example, a GIS can quickly generate a map with isopleth or contour lines that indicate differing amounts of rainfall. Such a map can be thought of as a rainfall contour map. Many sophisticated methods can estimate the characteristics of surfaces from a limited number of point measurements. A two-dimensional contour map created from the surface modelling of rainfall point measurements may be overlaid and analysed with any other map in a GIS covering the same area.

Additionally, from a series of three-dimensional points, or digital elevation model, isopleth lines representing elevation contours can be generated, along with slope analysis, shaded relief, and other elevation products. Watersheds can be easily defined for any given reach, by computing all of the areas contiguous and uphill from any given point of interest. Similarly, an expected thalweg of where surface water would want to travel in intermittent and permanent streams can be computed from elevation data in the GIS.

Topological Modelling

A GIS can recognize and analyse the spatial relationships that exist within digitally stored spatial data. These topological relationships allow complex spatial modelling and analysis to be performed. Topological relationships between geometric entities traditionally include adjacency (what adjoins what), containment (what encloses what), and proximity (how close something is to something else).

Networks

If all the factories near a wetland were accidentally to release chemicals into the river at the same time, how long would it take for a damaging amount of pollutant to enter the wetland reserve? A GIS can simulate the routing of materials along a linear network. Values such as slope, speed limit, or pipe diameter can be incorporated into network modelling to represent the flow of the phenomenon more accurately. Network modelling is commonly employed in transportation planning, hydrology modelling, and infrastructure modelling.

Hydrological Modelling

GIS hydrological models can provide a spatial element that other hydrological models lack, with the analysis of variables such as slope, aspect and watershed or catchment area. Terrain analysis is fundamental to hydrology, since water always flows down a slope. As basic terrain analysis of a DEM involves calculation of slope and aspect, DEMs are very useful for hydrological analysis. Slope and aspect can then be used to determine direction of surface runoff, and hence flow accumulation for the formation of streams, rivers and lakes. Areas of divergent flow can also give a clear indication of the boundaries of a catchment. Once a flow direction and accumulation matrix has been created, queries can be performed that show contributing or dispersal areas at a certain point. More detail can be added to the model, such as terrain roughness, vegetation types and soil types, which can influence infiltration and evapotranspiration rates, and hence influencing surface flow. These extra layers of detail ensures a more accurate model.

Cartographic Modelling

The term "cartographic modelling" was (probably) coined by Dana Tomlin in his PhD dissertation and later in his book which has the term in the title. Cartographic modelling refers to a process where several thematic layers of the same area are produced, processed, and analysed. Tomlin used raster layers, but the overlay method can be used more generally. Operations on map layers can be combined into algorithms, and eventually into simulation or optimization models.

Map Overlay

The combination of several spatial datasets (points, lines or polygons) creates a new output vector dataset, visually similar to stacking several maps of the same region. These overlays are similar to mathematical Venn diagram overlays. A union overlay combines the geographic features and attribute tables of both inputs into a single new output. An intersect overlay defines the area where both inputs overlap and retains a set of attribute fields for each. A symmetric difference overlay defines an output area that includes the total area of both inputs except for the overlapping area. Data extraction is a GIS process similar to vector overlay, though it can be used in either vector or raster data analysis. Rather than combining the properties and features of both datasets, data extraction involves using a "clip" or "mask" to extract the features of one data set that fall within the spatial extent of another dataset. In raster data analysis, the overlay of datasets is accomplished through a process known as "local operation on multiple rasters" or "map algebra," through a function that combines the values of each raster's matrix. This function may weigh some inputs more than others through use of an "index model" that reflects the influence of various factors upon a geographic phenomenon.

Automated Cartography

Digital cartography and GIS both encode spatial relationships in structured formal representations. GIS is used in digital cartography modelling as a (semi) automated process of making maps, so called Automated Cartography. In practice, it can be a subset of a GIS, within which it is equivalent to the stage of visualization, since in most cases not all of the GIS functionality is used. Cartographic products can be either in a digital or in a hardcopy format. Powerful analysis techniques with different data representation can produce high-quality maps within a short time period. The main problem in Automated Cartography is to use a single set of data to produce multiple products at a variety of scales, a technique known as cartographic generalization.

Geostatistics

Geostatistics is a branch of statistics focusing on spatial or spatiotemporal datasets. Developed originally to predict probability distributions of ore grades for mining operations, it is currently applied in diverse disciplines including petroleum geology, hydrogeology, hydrology, meteorology, oceanography, geochemistry, geometallurgy, geography, forestry, environmental control, landscape ecology, soil science, and agriculture (esp. in precision farming). Geostatistics is applied in varied branches of geography, particularly those involving the spread of diseases (epidemiology), the practice of commerce and military planning (logistics), and the development of efficient spatial networks. Geostatistical algorithms are incorporated in many places, including geographic information systems (GIS) and the R statistical environment.

Background

Geostatistics is intimately related to interpolation methods, but extends far beyond simple interpolation problems. It consists of a collection of numerical and mathematical techniques dealing with the characterization of spatial phenomena. Geostatistical techniques rely on statistical model that is based on random function (or random variable) theory to model the uncertainty associated with spatial estimation.

A number of simpler interpolation methods/algorithms, such as inverse distance weighting, linear interpolation and nearest-neighbour interpolation, were already well known before geostatistics. Geostatistics goes beyond the interpolation problem by considering the studied phenomenon at unknown locations as a random variable. Let $Z(x)$ be the value of the variable of interest at certain location.

This value is unknown (e.g. temperature, rainfall, piezometric level, geological facies, etc). Although there exists a value at location x that could be measured, geostatistics consider this value as random since it was not measured, or has not been measured yet. However, the randomness of $Z(x)$ is not complete, but defined by a cumulative distribution function (cdf) that depends on certain information that is known about the value $Z(x)$:

$$F(z, x) = \Pr ob\{Z(x) \leq | \text{ information}\}.$$

Typically, if the value of Z is known at locations close to x (or in the neighbourhood of x) one can constrain the pdf of $Z(x)$ to this neighbourhood. If a high spatial continuity is assumed, $Z(x)$ can only have values similar to the ones found in the neighbourhood. Conversely, in the absence of spatial continuity $Z(x)$ can take any value. The spatial continuity of the random variables is described by a model of spatial continuity that can be either a parametric function in the case of variogram-based geostatistics, or have a non-parametric form when using other methods such as multiple-point simulation or pseudo-genetic techniques.

By applying a single spatial model on an entire domain, one makes the assumption that Z is a stationary process. It means that the same statistical properties are applicable on the entire domain. Several geostatistical methods provide ways of relaxing this stationarity assumption.

In this framework, one can distinguish two modelling goals:

- Estimating the most probable value for $Z(x)$, which is equivalent to finding the mode of the pdf $f(z,x)$. This is usually denoted as an estimation problem.
- Working with the entire probability density function $f(z,x)$ by actually considering each possible outcome of it at each location. This is generally done by creating several alternative maps of Z, called realizations. Consider a domain discretized in N grid nodes (or pixels). Each realization is a sample of the complete N-dimensional joint distribution function

$$F(z, x) = \Pr ob\{Z(x_1) \leq z_1, Z(x_2) \leq z_2, \ldots, z(x_N) \leq z_N\}.$$

In this approach, the presence of multiple solutions to the interpolation problem is acknowledged. Each realization is considered as a possible scenario of what the real variable could be. All associated workflows are then considering ensemble of realizations, and consequently ensemble of predictions that allow for probabilistic forecasting. A number of methods exist for both geostatistical estimation and multiple realizations approaches. Several reference books provide a comprehensive overview of the discipline.

Kriging

Kriging is a group of geostatistical techniques to interpolate the value of a random field (e.g., the elevation, *z*, of the landscape as a function of the geographic location) at an unobserved location from observations of its value at nearby locations.

The theory behind interpolation and extrapolation by kriging was developed by the French mathematician Georges Matheron based on the Master's thesis of Daniel Gerhardus Krige, the pioneering plotter of distance-weighted average gold grades at the Witwatersrand reef complex in South Africa. The English verb is *to krige* and the most common noun is *kriging*; both are often pronounced with a hard "g", following the pronunciation of the name "Krige".

Applications of Kriging

The application of kriging to problems in geology and mining as well as to hydrology started in the mid-60s and especially in the 70s with the work of Georges Matheron. The connection between kriging and geostatistics remains prevalent today.

Kriging has been used in:

- Black box modelling in computer experiments
- Environmental science
- Hydrogeology
- Mining
- Natural resources
- Remote sensing.

Related Terms and Techniques

Kriging Terms

A series of related terms were also named after Krige, including kriged estimate, kriged estimator, kriging variance, kriging covariance, zero kriging variance, unity kriging covariance, kriging matrix, kriging method, kriging model, kriging plan, kriging process, kriging system, block kriging, co-kriging, disjunctive kriging, linear kriging, ordinary kriging, point kriging, random kriging, regular grid kriging, simple kriging and universal kriging.

Related Methods

Kriging is mathematically closely related to regression analysis. Both theories derive a best linear unbiased estimator, based on assumptions on covariances, make use of Gauss-Markov theorem to prove independence of the estimate and error, and make use of very similar formulae. They are nevertheless useful in different frameworks: kriging is made for interpolation of a single realisation of a random field, while regression models are based on multiple observations of a multivariate dataset.

In the statistical community the same technique is also known as *Gaussian process regression*, *Kolmogorov Wiener prediction*, or *best linear unbiased prediction*.

The kriging interpolation may also be seen as a spline in a reproducing kernel Hilbert space, with reproducing kernel given by the covariance function. The difference with the classical kriging approach is provided by the interpretation: while the spline is motivated by a minimum norm interpolation based on a Hilbert space structure, kriging is motivated by an expected squared prediction error based on a stochastic model.

Kriging with *polynomial trend surfaces* is mathematically identical to generalized least squares polynomial curve fitting.

Kriging can also be understood as a form of Bayesian inference. Kriging starts with a prior distribution over functions. This prior takes the form of a Gaussian process: N samples from a function will be normally distributed, where the covariance between any two samples is the covariance function (or kernel) of the Gaussian process evaluated at the spatial location of two points.

A set of values is then observed, each value associated with a spatial location. Now, a new value can be predicted at any new spatial location, by combining the Gaussian prior with a Gaussian likelihood function for each of the observed values. The resulting posterior distribution is also Gaussian, with a mean and covariance that can be simply computed from the observed values, their variance, and the kernel matrix derived from the prior.

History

The theory of kriging was developed by the French mathematician Georges Matheron based on the Master's thesis of Daniel Gerhardus Krige, the pioneering plotter of distance-weighted average gold grades at the Witwatersrand reef complex. The English verb is *to krige* and the most common adjective is *kriging*. The method was called *krigeage* for the first time in Matheron's 1960 *Krigeage d'un Panneau Rectangulaire par sa Peripherie*. Matheron, in this *Note Geostatistique No 28*, derives k^*, his *estimateur* and a precursor to the *kriged estimate* or *kriged estimator*. In classical statistics, Matheron's k^* is the length-weighted average grade of each of his *panneaux* in his set. What Matheron failed to derive was $var(k^*)$, the variance of his *estimateur*. On the contrary, he computed the length-weighted average grade of each *panneau* but did not compute the variance of its central value. In time, he replaced length-weighted average grades for three-dimensional sample spaces such as Matheronian blocks of ore with more abundant distance-weighted average grades for zero-dimensional sample spaces such as Matheronian points.

Multiple-indicator Kriging

Multiple-indicator kriging (MIK) is a recent advance on other techniques for mineral deposit modelling and resource block model estimation, such as ordinary kriging. Initially, *MIK* showed considerable promise as a new method that could more accurately estimate overall global mineral deposit concentrations or grades.

Problems

Unfortunately, the technique has not lived up to the promised benefits. Many experimental reviews over a range of actual mineral deposits has demonstrated that the technique has frequently been responsible for the over-estimation of reported resources. Typically observed with the technique is an inflation of the reported total tonnage of mineralized deposit material, occasionally along with a coincidental non-proportional reported deposit grade. This problem has been reported for a range of different deposit types, including most typical precious metals and base metals deposits.

The primary reason for the failure of the *MIK* technique has been demonstrated to be in part or totally due to each of the following deficiencies:

- The use of inherently large block sizes (required for adequate spatially distributed drill-hole and block sample support) does not allow for adequate resolution of geological geometry;
- The considerable level of difficulty in generating indicator cut-off variograms for all indicator cut-off grade levels;
- The failure of being able to determine the appropriate levels of post interpolation "uniform conditioning" support or adjustment for reported block grades;
- The allocating of probability block proportions for each indicator cut-off levels is difficult to apply;
- The accuracy of reported block grades often cannot be verified without the use of localized high density sampling.

Markov Chain

A Markov chain is a random process with the Markov property, i.e. the property, simply said, that the next state depends only on the current state and not on the past. It is a Markov model, named for Andrey Markov, for a particular type of Markov process in which the process can only be in a finite or countable number of states. Markov chains are useful as tools for statistical modelling in almost all fields of modern applied mathematics.

Introduction

Formally, a Markov chain is a discrete (discrete-time) random process with the Markov property. Often, the term "Markov chain" is used to mean a Markov process which has a discrete (finite or countable) state-space. Usually a Markov chain would be defined for a discrete set of times (i.e. a discrete-time Markov chain) although some authors use the same terminology where "time" can take continuous values. The use of the term in Markov chain Monte Carlo methodology covers cases where the process is in discrete-time (discrete algorithm steps) with a continuous state space. The following concentrates on the discrete-time discrete-state-space case.

A "discrete-time" random process means a system which is in a certain state at each "step", with the state changing randomly between steps. The steps are often thought of as time, but they can equally well refer to physical distance or any other discrete measurement; formally, the steps are just the integers or natural numbers, and the random process is a mapping of these to states. The Markov property states that the conditional probability distribution for the system at the next step (and in fact at all future steps) *given* its current state depends only on the current state of the system, and not additionally on the state of the system at previous steps.

Since the system changes randomly, it is generally impossible to predict the exact state of the system in the future. However, the statistical properties of the system's future can be predicted. In many applications it is these statistical properties that are important.

The changes of state of the system are called transitions, and the probabilities associated with various state-changes are called transition probabilities. The set of all states and transition probabilities completely characterizes a Markov chain. By convention, we assume all possible states and transitions have been included in the definition of the processes, so there is always a next-state and the process goes on forever.

A famous Markov chain is the so-called "drunkard's walk", a random walk on the number line where, at each step, the position may change by +1 or "1 with equal probability. From any position there are two possible transitions, to the next or previous integer. The transition probabilities depend only on the current position, not on the way the position was reached. For example, the transition probabilities from 5 to 4 and 5 to 6 are both 0.5, and all other transition probabilities from 5 are 0. These probabilities are independent of whether the system was previously in 4 or 6.

Another example is the dietary habits of a creature who eats only grapes, cheese or lettuce, and whose dietary habits conform to the following (artificial) rules: it eats exactly once a day; if it ate cheese yesterday, it will not today, and it will eat lettuce or grapes with equal probability; if it ate grapes yesterday, it will eat grapes today with probability 1/10, cheese with probability 4/10 and lettuce with probability 5/10; finally, if it ate lettuce yesterday, it won't eat lettuce again today but will eat grapes with probability 4/10 or cheese with probability 6/10. This creature's eating habits can be modelled with a Markov chain since its choice depends on what it ate yesterday, not additionally on what it ate 2 or 3 (or 4, etc.) days ago. One statistical property one could calculate is the expected percentage of the time the creature will eat grapes over a long period.

A series of independent events—for example, a series of coin flips—does satisfy the formal definition of a Markov chain. However, the theory is usually applied only when the probability distribution of the next step depends non-trivially on the current state.

Information Sciences

Markov chains are used throughout information processing. Claude Shannon's famous 1948 paper *A mathematical theory of communication*, which in a single step created the field of information theory, opens by introducing the concept of entropy through Markov modelling of the English language. Such idealized models can capture many of the statistical regularities of systems.

Even without describing the full structure of the system perfectly, such signal models can make possible very effective data compression through entropy encoding techniques such as arithmetic coding. They also allow effective state estimation and pattern recognition. Markov chains are also the basis for Hidden Markov Models, which are an important tool in such diverse fields as telephone networks (for error correction), speech recognition and bioinformatics. The world's mobile telephone systems depend on the Viterbi algorithm for error-correction, while hidden Markov models are extensively used in speech recognition and also in bioinformatics, for instance for coding region/gene prediction. Markov chains also play an important role in reinforcement learning.

Queueing Theory

Markov chains are the basis for the analytical treatment of queues (queueing theory). This makes them critical for optimizing the performance of telecommunications networks, where messages must often compete for limited resources (such as bandwidth).

Statistics

Markov chain methods have also become very important for generating sequences of random numbers to accurately reflect very complicated desired probability distributions, via a process called Markov chain Monte Carlo (MCMC). In recent years this has revolutionized the practicability of Bayesian inference methods, allowing a wide range of posterior distributions to be simulated and their parameters found numerically.

Economics and Finance

Markov chains are used in Finance and Economics to model a variety of different phenomena, including asset prices and market crashes. The first financial model to use a Markov chain was from Prasad *et al.* in 1974. Another was the regime-switching model of James D. Hamilton (1989), in which a Markov chain is used to model switches between periods of high volatility and low volatility of asset returns. A more recent example is the Markov Switching Multifractal asset pricing model, which builds upon the convenience of earlier regime-switching models. It uses an arbitrarily large Markov chain to drive the level of volatility of asset returns.

Dynamic macroeconomics heavily uses Markov chains. An example is using Markov chains to exogenously model prices of equity (stock) in a general equilibrium setting.

Leontief's Input-output model is a Markov chain.

Social Sciences

Markov chains are generally used in describing path-dependent arguments, where current structural configurations condition future outcomes. An example is the commonly argued link between economic development and the rise of capitalism. Once a country reaches a specific level of economic development, the configuration of structural factors, such as size of the commercial bourgeoisie, the ratio of urban to rural residence, the rate of political mobilization, etc., will generate a higher probability of transitioning from authoritarian to capitalist.

Mathematical Biology

Markov chains also have many applications in biological modelling, particularly population processes, which are useful in modelling processes that are (at least) analogous to biological populations. The Leslie matrix is one such example, though some of its entries are not probabilities (they may be greater than 1). Another example is the modelling of cell shape in dividing sheets of epithelial cells. Yet another example is the state of Ion channels in cell membranes.

Markov chains are also used in simulations of brain function, such as the simulation of the mammalian neocortex.

Gambling

Markov chains can be used to model many games of chance. The children's games Snakes and Ladders and "Hi Ho! Cherry-O", for example, are represented exactly by Markov chains. At each turn, the player starts in a given state (on a given square) and from there has fixed odds of moving to certain other states (squares).

Baseball

Markov chain models have been used in advanced baseball analysis since 1960, although their use is still rare. Each half-inning of a baseball game fits the Markov chain state when the number of runners and outs are considered. During any at-bat, there are 24 possible combinations of number of outs and position of the runners. Mark Pankin shows that Markov chain models can be used to evaluate runs created for both individual players as well as a team. The author also discusses various kinds of strategies and play conditions how Markov chain models have been used to analyse statistics for game situations such as bunting and base stealing and differences when playing on grass vs. astroturf.

Markov Text Generators

Markov processes can also be used to generate superficially "real-looking" text given a sample document: they are used in a variety of recreational "parody generator" software.

These processes are also used by spammers to inject real-looking hidden paragraphs into unsolicited email in an attempt to get these messages past spam filters.

History

Andrey Markov produced the first results (1906) for these processes, purely theoretically. A generalization to countably infinite state spaces was given by Kolmogorov (1936). Markov chains are related to Brownian motion and the ergodic hypothesis, two topics in physics which were important in the early years of the twentieth century, but Markov appears to have pursued this out of a mathematical motivation, namely the extension of the law of large numbers to dependent events. In 1913, he applied his findings for the first time to the first 20,000 letters of Pushkin's *Eugene Onegin.*

Seneta provides an account of Markov's motivations and the theory's early development. The term "chain" was used by Markov (1906).

Markov Chain Geostatistics

Markov chain geostatistics refer to the Markov chain models, simulation algorithms and associated spatial correlation measures (e.g., transiogram) based on the Markov chain random field theory, which extends a single Markov chain into a multi-dimensional field for geostatistical modelling. A Markov chain random field is still a single spatial Markov chain. The spatial Markov chain moves or jumps in a space and decides its state at a location through interactions with its nearest known neighbours in different directions, including its last stay location. Because single-step transition probability matrices are difficult to estimate from sparse sample data and are impractical in representing the complex spatial heterogeneity of states, the transiogram, which is defined as a transition probability function over the distance lag, is proposed as the accompanying spatial measure of Markov chain random fields.

Support Vector Machine

Support vector machines (SVMs) are a set of related supervised learning methods that analyse data and recognize patterns, used for classification and regression analysis. The original SVM algorithm was invented by Vladimir Vapnik and the current standard incarnation (soft margin) was proposed by Corinna Cortes and Vladimir Vapnik.

The standard SVM takes a set of input data, and predicts, for each given input, which of two possible classes the input is a member of, which makes the SVM a non-probabilistic binary linear classifier. Since an SVM is a classifier, then given a set of training examples, each marked as belonging to one of two categories, an SVM training algorithm builds a model that predicts whether a new example falls into one category or the other. Intuitively, an SVM model is a representation of the examples as points in space, mapped so that the examples of the separate categories are divided by a clear gap that is as wide as possible. New examples are then mapped into that same space and predicted to belong to a category based on which side of the gap they fall on.

More formally, a support vector machine constructs a hyperplane or set of hyperplanes in a high or infinite dimensional space, which can be used for classification, regression or other tasks. Intuitively, a good separation is achieved by the hyperplane that has the largest distance to the nearest training data points of any class (so-called functional margin), since in general the larger the margin the lower the generalization error of the classifier.

Whereas the original problem may be stated in a finite dimensional space, it often happens that in that space the sets to be discriminated are not linearly separable. For this reason it was proposed that the original finite dimensional space be mapped into a much higher dimensional space presumably making the separation easier in that space. SVM schemes use a mapping into a larger space so that cross products may be computed easily in terms of the variables in the original space making the computational load reasonable. The cross products in the larger space are defined in terms of a kernel function $K(x,y)$ which can be selected to suit the problem. The hyperplanes in the large space are defined as the set of points whose cross product with a vector in that space is constant. The vectors defining the hyperplanes can be chosen to be linear combinations with parameters α_i of images of feature vectors which occur in the data base.

Motivation

Classifying data is a common task in machine learning. Suppose some given data points each belong to one of two classes, and the goal is to decide which class a *new* data point will be in. In the case of support vector machines, a data point is viewed as a p-dimensional vector (a list of p numbers), and we want to know whether we can separate such points with a $(p - 1)$-dimensional hyperplane. This is called a linear classifier. There are many hyperplanes that might classify the data. One reasonable choice as the best hyperplane is the one that represents the largest separation, or margin, between the two classes. So we choose the hyperplane so that the distance from it to the nearest data point on each side is maximized. If such a hyperplane exists, it is known as the *maximum-margin hyperplane* and the linear classifier it defines is known as a *maximum margin classifier*.

Multiclass SVM

Multiclass SVM aims to assign labels to instances by using support vector machines, where the labels are drawn from a finite set of several elements. The dominating approach for doing so is to reduce the single multiclass problem into multiple binary classification problems. Each of the problems yields a binary classifier, which is assumed to produce an output function that gives relatively large values for examples from the positive class and relatively small values for examples belonging to the negative class. Two common methods to build such binary classifiers are where each classifier distinguishes between (i) one of the labels to the rest (one-versus-all) or (ii) between every pair of classes (one-versus-one). Classification of new instances for one-versus-all case is done by a winner-takes-all strategy, in which the classifier with the highest output function assigns the class (it is important that the output functions be calibrated to produce comparable scores). For the one-versus-one approach, classification is done by a max-wins voting strategy, in which every classifier assigns the instance to one of the two classes, then the vote for the assigned class is increased by one vote, and finally the class with most votes determines the instance classification.

Structured SVM

SVMs have been generalized to structured SVMs, where the label space is structured and of possibly infinite size.

Regression

A version of SVM for regression was proposed in 1996 by Vladimir Vapnik, Harris Drucker, Chris Burges, Linda Kaufman and Alex Smola. This method is called support vector regression (SVR). The model produced by support vector classification (as described above) depends only on a subset of the training data, because the cost function for building the model does not care about training points that lie beyond the margin. Analogously, the model produced by SVR depends only on a subset of the training data, because the cost function for building the model ignores any training data close to the model prediction (within a threshold ε). There is also a least squares version of support vector machine (SVM) called least squares support vector machine (LS-SVM) proposed in Suykens and Vandewalle.

Implementation

The parameters of the maximum-margin hyperplane are derived by solving the optimization. There exist several specialized algorithms for quickly solving the QP problem that arises from SVMs, mostly reliant on heuristics for breaking the problem down into smaller, more-manageable chunks. A common method for solving the QP problem is Platt's Sequential Minimal Optimization (SMO) algorithm, which breaks the problem down into 2-dimensional sub-problems that may be solved analytically, eliminating the need for a numerical optimization algorithm.

Another approach is to use an interior point method that uses Newton-like iterations to find a solution of the Karush–Kuhn–Tucker conditions of the primal and dual problems. Instead of solving a sequence of broken down problems, this approach directly solves the problem as a whole. To avoid solving a linear system involving the large kernel matrix, a low rank approximation to the matrix is often used to use the kernel trick.

Cellular Automaton

A cellular automaton (pl. cellular automata, abbrev. CA) is a discrete model studied in computability theory, mathematics, physics, complexity science, theoretical biology and microstructure modelling. It consists of a regular grid of *cells*, each in one of a finite number of *states*, such as "On" and "Off" (in contrast to a coupled map lattice). The grid can be in any finite number of dimensions. For each cell, a set of cells called its *neighbourhood* (usually including the cell itself) is defined relative to the specified cell. For example, the neighbourhood of a cell might be defined as the set of cells a distance of 2 or less from the cell. An initial state (time t=0) is selected by assigning a state for each cell. A new *generation* is created (advancing t by 1), according to some fixed rule (generally, a mathematical function) that determines the new state of each cell in terms of the current state of the cell and the states of the cells in its neighbourhood. For example, the rule might be that the cell is "On" in the next generation if exactly two of the cells in the neighbourhood are "On" in the current generation, otherwise the cell is "Off" in the next generation. Typically, the rule for updating the state of cells is the same for each cell and does not change over time, and is applied to the whole grid simultaneously, though exceptions are known.

Overview

One way to simulate a two-dimensional cellular automaton is with an infinite sheet of graph paper along with a set of rules for the cells to follow. Each square is called a "cell" and each cell has two possible states, black and white. The "neighbours" of a cell are the 8 squares touching it. For such a cell and its neighbours, there are 512 (= 2^9) possible patterns. For each of the 512 possible patterns, the rule table would state whether the centre cell will be black or white on the next time interval. Conway's Game of Life is a popular version of this model.

It is usually assumed that every cell in the universe starts in the same state, except for a finite number of cells in other states, often called a *configuration*. More generally, it is sometimes assumed that the universe starts out covered with a periodic pattern, and only a finite number of cells violate that pattern. The latter assumption is common in one-dimensional cellular automata.

Cellular automata are often simulated on a finite grid rather than an infinite one. In two dimensions, the universe would be a rectangle instead of an infinite plane. The obvious problem with finite grids is how to handle the cells on the edges. How they are handled will affect the values of all the cells in the grid. One possible method is to allow the values in those cells to remain constant. Another method is to define neighbourhoods differently for these cells. One could say that they have fewer neighbours, but then one would also have to define new rules for the cells located on the edges. These cells are usually handled with a *toroidal* arrangement: when one goes off the top, one comes in at the corresponding position on the bottom, and when one goes off the left, one comes in on the right. (This essentially simulates an infinite periodic tiling, and in the field of partial differential equations is sometimes referred to as *periodic* boundary conditions.) This can be visualized as taping the left and right edges of the rectangle to form a tube, then taping the top and bottom edges of the tube to form a torus (doughnut shape). Universes of other dimensions are handled similarly. This is done in order to solve boundary problems with neighbourhoods, but another advantage of this system is that it is easily programmable using modular arithmetic functions. For example, in a 1-dimensional cellular automaton like the examples below, the neighbourhood of a cell x_i^t—where t is the time step (vertical), and i is the index (horizontal) in one generation—is $\{x_{i-1}^{t-1}, x_i^{t-1}, x_{i+1}^{t-1}\}$. There will obviously be problems when a neighbourhood on a left border references its upper left cell, which is not in the cellular space, as part of its neighbourhood.

History

Stanishaw Ulam, while working at the Los Alamos National Laboratory in the 1940s, studied the growth of crystals, using a simple lattice network as his model. At the same time, John von Neumann, Ulam's colleague at Los Alamos, was working on the problem of self-replicating systems. Von Neumann's initial design was founded upon the notion of one robot building another robot. This design is known as the kinematic model. As he developed this design, von Neumann came to realize the great difficulty of building a self-replicating robot, and of the great cost in providing the robot with a "sea of parts" from which to build its replicant. Ulam suggested that von Neumann develop his design around a mathematical abstraction, such as the one Ulam used to study crystal growth. Thus was born the first system of cellular automata. Like Ulam's lattice network, von Neumann's cellular automata are two-dimensional, with his self-replicator implemented algorithmically. The result was a universal copier and constructor working within a CA with a small neighbourhood (only those cells that touch are neighbours; for von Neumann's cellular automata, only orthogonal cells), and with 29 states per cell. Von Neumann gave an existence proof that a particular pattern would make endless copies of itself within the given cellular universe. This design is known as the tessellation model, and is called a von Neumann universal constructor.

Also in the 1940s, Norbert Wiener and Arturo Rosenblueth developed a cellular automaton model of excitable media. Their specific motivation was the mathematical description of impulse conduction in cardiac systems. Their original work continues to be cited in modern research publications on cardiac arrhythmia and excitable systems.

In the 1960s, cellular automata were studied as a particular type of dynamical system and the connection with the mathematical field of symbolic dynamics was established for the first time. In 1969, G. A. Hedlund compiled many results following this point of view in what is still considered as a seminal paper for the mathematical study of cellular automata. The most fundamental result is the characterization of the set of global rules of cellular automata as the set of continuous endomorphisms of shift spaces.

In the 1970s a two-state, two-dimensional cellular automaton named Game of Life became very widely known, particularly among the early computing community. Invented by John Conway and popularized by Martin Gardner in a *Scientific American* article, its rules are as follows: If a cell has 2 black neighbours, it stays the same. If it has 3 black neighbours, it becomes black. In all other situations it becomes white. Despite its simplicity, the system achieves an impressive diversity of Behaviour, fluctuating between apparent randomness and order. One of the most apparent features of the Game of Life is the frequent occurrence of *gliders*, arrangements of cells that essentially move themselves across the grid. It is possible to arrange the automaton so that the gliders interact to perform computations, and after much effort it has been shown that the Game of Life can emulate a universal Turing machine. Possibly because it was viewed as a largely recreational topic, little follow-up work was done outside of investigating the particularities of the Game of Life and a few related rules.

In 1969, however, German computer pioneer Konrad Zuse published his book *Calculating Space*, proposing that the physical laws of the universe are discrete by nature, and that the entire universe is the output of a deterministic computation on a giant cellular automaton. This was the first book on what today is called digital physics.

In 1983 Stephen Wolfram published the first of a series of papers systematically investigating a very basic but essentially unknown class of cellular automata, which he terms *elementary cellular automata.* The unexpected complexity of the Behaviour of these simple rules led Wolfram to suspect that complexity in nature may be due to similar mechanisms. Additionally, during this period Wolfram formulated the concepts of intrinsic randomness and computational irreducibility, and suggested that rule 110 may be universal—a fact proved later by Wolfram's research assistant Matthew Cook in the 1990s.

In 2002 Wolfram published a 1280-page text *A New Kind of Science,* which extensively argues that the discoveries about cellular automata are not isolated facts but are robust and have significance for all disciplines of science. Despite much confusion in the press and academia, the book did not argue for a fundamental theory of physics based on cellular automata, and although it did describe a few specific physical models based on cellular automata, it also provided models based on qualitatively different abstract systems.

Elementary Cellular Automata

The simplest nontrivial CA would be one-dimensional, with two possible states per cell, and a cell's neighbours defined to be the adjacent cells on either side of it. A cell and its two neighbours form a neighbourhood of 3 cells, so there are $2^3=8$ possible patterns for a neighbourhood. A rule consists of deciding, for each pattern, whether the cell will be a 1 or a 0 in the next generation. There are then $2^8=256$ possible rules. These 256 CAs are generally referred to by their Wolfram code, a standard naming convention invented by Stephen Wolfram which gives each rule a number from 0 to 255. A number of papers have analysed and compared these 256 CAs. The rule 30 and rule 110 CAs are particularly interesting. The images below show the history of each when the starting configuration consists of a 1 (in the centre of each image) surrounded by 0's. Each row of pixels represents a generation in the history of the automation, with t=0 being the top row. Each pixel is coloured white for 0 and black for 1.

Rule 30 generates exhibits *class 3* Behaviour, meaning even simple input patterns such as that shown lead to chaotic, seemingly random histories.

Rule 110, like the Game of Life, exhibits what Wolfram calls *class 4* Behaviour, which is neither completely random nor completely repetitive. Localized structures appear and interact in various complicated-looking ways. In the course of the development of *A New Kind of Science*, as a research assistant to Stephen Wolfram in 1994, Matthew Cook proved that some of these structures were rich enough to support universality. This result is interesting because rule 110 is an extremely simple one-dimensional system, and one which is difficult to engineer to perform specific Behaviour. This result therefore provides significant support for Wolfram's view that class 4 systems are inherently likely to be universal. Cook presented his proof at a Santa Fe Institute conference on Cellular Automata in 1998, but Wolfram blocked the proof from being included in the conference proceedings, as Wolfram did not want the proof to be announced before the publication of *A New Kind of Science*. In 2004, Cook's proof was finally published in Wolfram's journal Complex Systems (Vol. 15, No. 1), over ten years after Cook came up with it. Rule 110 has been the basis over which some of the smallest universal Turing machines have been built, inspired on the breakthrough concepts that the development of the proof of rule 110 universality produced.

Reversible

A CA is said to be *reversible* if for every current configuration of the CA there is exactly one past configuration (preimage). If one thinks of a CA as a function mapping configurations to configurations, reversibility implies that this function is objective.

For one dimensional CA there are known algorithms for deciding whether a rule is reversible or irreversible. For CA of two or more dimensions it has been proved that the reversibility is undecidable for arbitrary rules. The proof by Jarkko Kari is related to the tiling problem by Wang tiles.

Reversible CA are often used to simulate such physical phenomena as gas and fluid dynamics, since they obey the laws of thermodynamics. Such CA have rules specially constructed to be reversible. Such systems have been studied by Tommaso Toffoli, Norman Margolus and others.

For finite CAs that are not reversible, there must exist patterns for which there are no previous states. These patterns are called *Garden of Eden patterns*. In other words, no pattern exists which will develop into a Garden of Eden pattern. Several techniques can be used to explicitly construct reversible CA with known inverses. Two common ones are the second order technique and the partitioning technique, both of which involve modifying the definition of a CA in some way. Although such automata do not strictly satisfy the definition given above, it can be shown that they can be emulated by conventional CAs with sufficiently large neighbourhoods and numbers of states, and can therefore be considered a subset of conventional CA.

Totalistic

A special class of CAs are *totalistic* CAs. The state of each cell in a totalistic CA is represented by a number (usually an integer value drawn from a finite set), and the value of a cell at time t depends only on the *sum* of the values of the cells in its neighbourhood (possibly including the cell itself) at time $t–1$. If the state of the cell at time t does depend on its own state at time $t–1$ then the CA is properly called *outer totalistic*, although the distinction is not always made. Conway's Game of Life is an example of an outer totalistic CA with cell values 0 and 1.

A notation exists to describe rulesets of two-state totalistic CAs consisting of an initial indicating the neighbourhood of each cell and sums following the letters S (for survival) and B (for birth) for which those changes occur. In this notation Conway's Game of Life is M:S23/B3. This notation has been extended for non-totalistic CAs, where a letter or letters follow each sum indicating what patterns of neighbours cause survival or birth events.

Wolfram Classes

Stephan Wolfram, in A New Kind of Science and in several papers dating from the mid-1980s, defined four classes into which cellular automata and several other simple computational models can be divided depending on their Behaviour. While earlier studies in cellular automata tended to try to identify type of patterns for specific rules, Wolfram's classification was the first attempt to classify the rules themselves. In order of complexity the classes are:

- Class 1: Nearly all initial patterns evolve quickly into a stable, homogeneous state. Any randomness in the initial pattern disappears.
- Class 2: Nearly all initial patterns evolve quickly into stable or oscillating structures. Some of the randomness in the initial pattern may filter out, but some remains. Local changes to the initial pattern tend to remain local.
- Class 3: Nearly all initial patterns evolve in a pseudo-random or chaotic manner. Any stable structures that appear are quickly destroyed by the surrounding noise. Local changes to the initial pattern tend to spread indefinitely.
- Class 4: Nearly all initial patterns evolve into structures that interact in complex and interesting ways. Class 2 type stable or oscillating structures may be the eventual outcome, but the number of steps required to reach this state may be very large, even when the initial pattern is relatively simple. Local changes to the initial pattern may spread indefinitely. Wolfram has conjectured that many, if not all class 4 cellular automata are capable of universal computation. This has been proven for Rule 110 and Conway's game of life.

These definitions are qualitative in nature and there is some room for interpretation. According to Wolfram, with almost any general classification scheme there are inevitably cases which get assigned to one class by one definition and another class by another definition. And so it is with cellular automata: there are occasionally rules...that show some features of one class and some of another.

Evolving Cellular Automata using Genetic Algorithms

Recently there has been a keen interest in building decentralized systems, be they sensor networks or more sophisticated micro level structures designed at the network level and aimed at decentralized information processing. The idea of emergent computation came from the need of using distributed systems to do information processing at the global level. The area is still in its infancy, but some people have started taking the idea seriously. Melanie Mitchell who is Professor of Computer Science at Portland State University and also the Santa Fe Institute External professor has been working on the idea of using self-evolving cellular arrays to study emergent computation and distributed information processing. Mitchell and colleagues are using evolutionary computation to program cellular arrays. Computation in decentralized systems is very different from classical systems, where the information is processed at some central location depending on the system's state. In decentralized system, the information processing occurs in the form of global and local pattern dynamics.

The inspiration for this approach comes from complex natural systems like insect colonies, nervous system and economic systems. The focus of the research is to understand how computation occurs in an evolving decentralized system. In order to model some of the features of these systems and study how they give rise to emergent computation, Mitchell and collaborators at the SFI have applied Genetic Algorithms to evolve patterns in cellular automata.

They have been able to show that the GA discovered rules that gave rise to sophisticated emergent computational strategies. Mitchell's group used a single dimensional binary array where each cell has two neighbours. The array can be thought of as a circle where the first and last cells are neighbours. The evolution of the array was tracked through the number of ones and zeros after each iteration. The results were plotted to show clearly how the network evolved and what sort of emergent computation was visible.

The results produced by Mitchell's group are interesting, in that a very simple array of cellular automata produced results showing coordination over global scale, fitting the idea of emergent computation. Future work in the area may include more sophisticated models using cellular automata of higher dimensions, which can be used to model complex natural systems.

Cryptography Use

Cellular automata have been proposed for public key cryptography. The one way function is the evolution of a finite CA whose inverse is believed to be hard to find. Given the rule, anyone can easily calculate future states, but it appears to be very difficult to calculate previous states. However, the designer of the rule can create it in such a way as to be able to easily invert it. Therefore, it is apparently a trapdoor function, and can be used as a public-key cryptosystem. The security of such systems is not currently known.

Related Automata

There are many possible generalizations of the CA concept.

One way is by using something other than a rectangular (cubic, *etc.*) grid. For example, if a plane is tiled with regular hexagons, those hexagons could be used as cells. In many cases the resulting cellular automata are equivalent to those with rectangular grids with specially designed neighbourhoods and rules.

Also, rules can be probabilistic rather than deterministic. A probabilistic rule gives, for each pattern at time t, the probabilities that the central cell will transition to each possible state at time $t+1$. Sometimes a simpler rule is used; for example: "The rule is the Game of Life, but on each time step there is a 0.001% probability that each cell will transition to the opposite colour."

The neighbourhood or rules could change over time or space. For example, initially the new state of a cell could be determined by the horizontally adjacent cells, but for the next generation the vertical cells would be used.

The grid can be finite, so that patterns can "fall off" the edge of the universe.

In CA, the new state of a cell is not affected by the new state of other cells. This could be changed so that, for instance, a 2 by 2 block of cells can be determined by itself and the cells adjacent to itself.

There are *continuous automata*. These are like totalistic CA, but instead of the rule and states being discrete, continuous functions are used, and the states become continuous (usually values in [0,1]). The state of a location is a finite number of real numbers. Certain CA can yield diffusion in liquid patterns in this way.

Continuous spatial automata have a continuum of locations. The state of a location is a finite number of real numbers. Time is also continuous, and the state evolves according to differential equations. One important example is reaction-diffusion textures, differential equations proposed by Alan Turing to explain how chemical reactions could create the stripes on zebras and spots on leopards. When these are approximated by CA, such CAs often yield similar patterns. MacLennan considers continuous spatial automata as a model of computation.

There are known examples of continuous spatial automata which exhibit propagating phenomena analogous to gliders in the Game of Life.

Natural Biotic Types

Some living things use naturally occurring cellular automata in their functioning.

Patterns of some seashells, like the ones in *Conus* and *Cymbiola* genus, are generated by natural CA. The pigment cells reside in a narrow band along the shell's lip. Each cell secretes pigments according to the activating and inhibiting activity of its neighbour pigment cells, obeying a natural version of a mathematical rule. The cell band leaves the coloured pattern on the shell as it grows slowly. For example, the widespread species *Conus textile* bears a pattern resembling the Rule 30 CA described above.

Plants regulate their intake and loss of gases via a CA mechanism. Each stoma on the leaf acts as a cell.

Neural networks can be used as cellular automata, too. The complex moving wave patterns on the skin of cephalopods are a good display of corresponding activation patterns in the animals' brain.

Chemical Types

The Belousov-Zhabotinsky reaction is a spatio-temporal chemical oscillator which can be simulated by means of a cellular automaton. In the 1950s A. M. Zhabotinsky (extending the work of B. P. Belousov) discovered that when a thin, homogenous layer of a mixture of malonic acid, acidified bromate and a ceric salt were mixed together and left undisturbed, fascinating geometric patterns such as concentric circles and spirals propagate across the medium. In the "Computer Recreations" section of the August 1988 issue of Scientific American, A. K. Dewdney discussed a cellular automaton which was developed by Martin Gerhardt and Heike Schuster of the University of Bielefeld (West Germany). This automaton produces wave patterns resembling those in the Belousov-Zhabotinsky reaction.

Computer Processors

CA processors are a physical, not software only, implementation of CA concepts, which can process information computationally. Processing elements are arranged in a regular grid of identical cells. The grid is usually a square tiling, or tessellation, of two or three dimensions; other tilings are possible, but not yet used. Cell states are determined only by interactions with the small number of adjoining cells. Cells interact, communicate, directly only with adjoining, adjacent, neighbour cells. No means exists to communicate directly with cells farther away. One such CA processor array configuration is the systolic array. Cell interaction can be via electric charge, magnetism, vibration (phonons at quantum scales), or any other physically useful means. This can be done in several ways so no wires are needed between any elements.

This is very unlike processors used in most computers today, von Neumann designs, which are divided into sections with elements that can communicate with distant elements, over wires.

Error Correction Coding

CA have been applied to design error correction codes in the paper "Design of CAECC – Cellular Automata Based Error Correcting Code", by D. Roy Chowdhury, S. Basu, I. Sen Gupta, P. Pal Chaudhuri. The paper defines a new scheme of building SEC-DED codes using CA, and also reports a fast hardware decoder for the code.

Regionalized Variable Theory

Regionalized Variable Theory (RVT) is a geostatistical method used for interpolation in space.

The concept of the theory is that interpolation from points in space should not be based on a smooth continuous object. It should be, however, based on a stochastic model that takes into consideration the various trends in the original set of points. The theory considers that within any dataset, three types of relationships can be detected:

1. Structural part, which is also called the trend.
2. Correlated variation.
3. Uncorrelated variation, or noise.

After defining the above three relationships, RVT then applies the first law of geography, in order to predict the unknown values of points. The major application of this theory is the Kriging method for interpolation.

Chapter 4

Earth Observation Satellite

Earth observation satellites are satellites specifically designed to observe Earth from orbit, similar to reconnaissance satellites but intended for non-military uses such as environmental monitoring, meteorology, map making etc. Geostationary satellites hover over the same spot, providing continuous monitoring to a portion of the Earth's surface. Polar orbiting satellites provide global coverage, but only twice per day at any given spot.

Weather Satellite

The weather satellite is a type of satellite that is primarily used to monitor the weather and climate of the Earth. Satellites can be either polar orbiting, seeing the same swath of the Earth every 12 hours, or geostationary, hovering over the same spot on Earth by orbiting over the equator while moving at the speed of the Earth's rotation. These meteorological satellites, however, see more than clouds and cloud systems. City lights, fires, effects of pollution, auroras, sand and dust storms, snow cover, ice mapping, boundaries of ocean currents, energy flows, etc., and other types of environmental information are collected using weather satellites. Weather satellite images helped in monitoring the volcanic ash cloud from Mount St. Helens and activity from other volcanoes such as Mount Etna. Smoke from fires in the western United States such as Colorado and Utah have also been monitored.

Other environmental satellites can detect changes in the Earth's vegetation, sea state, ocean colour, and ice fields. For example, the 2002 oil spill off the northwest coast of Spain was watched carefully by the European ENVISAT, which, though not a weather satellite, flies an instrument (ASAR) which can see changes in the sea surface. El Nino and its effects on weather are monitored daily from satellite images. The Antarctic ozone hole is mapped from weather satellite data. Collectively, weather satellites flown by the U.S., Europe, India, China, Russia, and Japan provide nearly continuous observations for a global weather watch.

Observation

Observation is typically made via different 'channels' of the Electromagnetic spectrum, in particular, the Visible and Infrared portions.

Some of these channels include :

- *Visible and Near Infrared:* 0.6 mm - 1.6 mm - For recording cloud cover during the day
- *Infrared:* 3.9 mm - 7.3 mm (Water Vapour), 8.7 mm, - 13.4 mm (Thermal imaging).

History

The first weather satellite, Vanguard 2, was launched on February 17, 1959. It was designed to measure cloud cover and resistance, but a poor axis of rotation kept it from collecting a notable amount of useful data. The first weather satellite to be considered a success was TIROS-1, launched by NASA on 1 April 1960. TIROS operated for 78 days and proved to be much more successful than Vanguard 2. TIROS paved the way for the Nimbus program, whose technology and findings are the heritage of most of the Earth-observing satellites NASA and NOAA have launched since then.

Visible Spectrum

Visible-light images from weather satellites during local daylight hours are easy to interpret even by the average person; clouds, cloud systems such as fronts and tropical storms, lakes, forests, mountains, snow ice, fires, and pollution such as smoke, smog, dust and haze are readily apparent. Even wind can be determined by cloud patterns, alignments and movement from successive photos.

Infrared Spectrum

The thermal or infrared images recorded by sensors called scanning radiometers enable a trained analyst to determine cloud heights and types, to calculate land and surface water temperatures, and to locate ocean surface features. Infrared satellite imagery can be used effectively for tropical cyclones with a visible eye pattern, using the Dvorak technique, where the difference between the temperature of the warm eye and the surrounding cold cloud tops can be used to determine its intensity (colder cloud tops generally indicate a more intense storm).

Infrared pictures depict ocean eddies or vortices and map currents such as the Gulf Stream which are valuable to the shipping industry. Fishermen and farmers are interested in knowing land and water temperatures to protect their crops against frost or increase their catch from the sea. Even El Nino phenomena can be spotted. Using colour-digitized techniques, the gray shaded thermal images can be converted to colour for easier identification of desired information.

Types

There are two basic types of meteorological satellites: geostationary and polar orbiting.

Geostationary

Geostationary weather satellites orbit the Earth above the equator at altitudes of 35,880 km (22,300 miles). Because of this orbit, they remain stationary with respect to the rotating Earth and thus can record or transmit images of the entire hemisphere below continuously with their visible-light and infrared sensors. The news media use the geostationary photos in their daily weather presentation as single images or made into movie loops. These are also available on the city forecast pages of noaa.gov (example Dallas, TX).

Several geostationary meteorological spacecraft are in operation. The United States has two in operation; GOES-11 and GOES-12. GOES-12, designated GOES-East, is located over the Amazon River and provides most of the U.S. weather information. GOES-11 is GOES-West over the eastern Pacific Ocean. The Japanese have one in operation; MTSAT-1R over the mid Pacific at 140°E. The Europeans have Meteosat-8 (3.5°W) and Meteosat-9 (0°) over the Atlantic Ocean and have Meteosat-6 (63°E) and Meteosat-7 (57.5°E) over the Indian Ocean. The Russians operate the GOMS over the equator south of Moscow. India also operates geostationary satellites called INSAT which carry instruments for meteorological purposes. China operated the geostationary satellites FY-2D at 86.5°E and FY-2E at 123.5°E, which are no longer in use anymore.

Polar Orbiting

Polar orbiting weather satellites circle the Earth at a typical altitude of 850 km (530 miles) in a north to south (or vice versa) path, passing over the poles in their continuous flight. Polar satellites are in sun-synchronous orbits, which means they are able to observe any place on Earth and will view every location twice each day with the same general lighting conditions due to the near-constant local solar time. Polar orbiting weather satellites offer a much better resolution than their geostationary counterparts due their closeness to the Earth.

The United States has the NOAA series of polar orbiting meteorological satellites, presently NOAA 17 and NOAA 18 as primary spacecraft, NOAA 15 and NOAA 16 as secondary spacecraft, NOAA 14 in standby, and NOAA 12. Europe has the Metop-A satellite. Russia has the Meteor and RESURS series of satellites. China has FY-1D and FY-3A. India has polar orbiting satellites as well.

DMSP

The United States Department of Defence's Meteorological Satellite (DMSP) can "see" the best of all weather vehicles with its ability to detect objects almost as 'small' as a huge oil tanker. In addition, of all the weather satellites in orbit, only DMSP can "see" at night in the visual. Some of the most spectacular photos have been recorded by the night visual sensor; city lights, volcanoes, fires, lightning, meteors, oil field burn-offs, as well as the Aurora Borealis and Aurora Australis have been captured by this 450-mile-high space vehicle's low moonlight sensor.

At the same time, energy monitoring as well as city growth can be accomplished since both major and even minor cities, as well as highway lights, are conspicuous. This informs astronomers of light pollution. The New York City Blackout of 1977 was captured by one of the night orbiter DMSP space vehicles.

In addition to monitoring city lights, these photos are a life saving asset in the detection and monitoring of fires. Not only do the satellites see the fires visually day and night, but the thermal and infrared scanners on board these weather satellites detect potential fire sources below the surface of the Earth where smoldering occurs. Once the fire is detected, the same weather satellites provide vital information about wind that could fan or spread the fires. These same cloud photos from space tell the firefighter when it will rain.

Dramatic photos are provided by all the weather satellites, but even more definitive were the DMSP night visible-light pictures of the 700 oil well fires that Iraq started on 23 February 1991 as they fled Kuwait. These fires were vividly illustrated as huge flashes in the night photos, far outstripping the glow of large populated areas. The fires consumed millions of gallons of oil; the last was doused on November 6.

Uses

Snowfield monitoring, especially in the Sierra Nevada, can be helpful to the hydrologist keeping track of how much snow is available for runoff vital to the water sheds of the western United States. This information is gleaned from existing satellites of all agencies of the U.S. government (in addition to local, on-the-ground measurements). Ice floes, packs and bergs can also be located and tracked from weather space craft.

Even pollution whether it's nature-made or man-made can be pinpointed. The visual and infrared photos show effects of pollution from their respective areas over the entire earth. Aircraft and rocket pollution, as well as condensation trails, can also be spotted.

The ocean current and low level wind information gleaned from the space photos can help predict oceanic oil spill coverage and movement. Almost every summer, sand and dust from the Sahara Desert in Africa drifts across the equatorial regions of the Atlantic Ocean. GOES-EAST photos enable meteorologists to observe, track and forecast this sand cloud.

In addition to reducing visibilities and causing respiratory problems, sand clouds suppress hurricane formation by modifying the solar radiation balance of the tropics. Other dust storms in Asia and mainland China are common and easy to spot and monitor, with recent examples of dust moving across the Pacific ocean and reaching North America.

In remote areas of the world with few local observers, fires could rage out of control for days or even weeks and consume millions of acres before authorities are alerted. Weather satellites can be a tremendous asset in such situations. Nighttime photos also clearly show the burn-off in the gas and oil fields of the Middle East and African countries. This burn-off throws large amounts of carbon dioxide into the atmosphere.

Environmental Monitoring

Other environmental satellites can assist environmental monitoring by detecting changes in the Earth's vegetation, sea state, ocean colour, and ice fields. By monitoring vegetation changes over time, droughts can be monitored by comparing the current vegetation state to its long term average. For example, the 2002 oil spill off the northwest coast of Spain was watched carefully by the European ENVISAT, which, though not a weather satellite, flies an instrument (ASAR) which can see changes in the sea surface.

Mapping

Terrain can be mapped from space with the use of satellites, such as RADARSAT-1 and TerraSAR-X.

Sonar

Sonar (originally an acronym for Sound Navigation And Ranging) is a technique that uses sound propagation (usually underwater, as in Submarine navigation) to navigate, communicate with or detect other vessels. Two types of technology share the name "sonar": *passive* sonar is essentially listening for the sound made by vessels; *active* sonar is emitting pulses of sounds and listening for echoes. Sonar may be used as a means of acoustic location and of measurement of the echo characteristics of "targets" in the water. Acoustic location in air was used before the introduction of radar. Sonar may also be used in air for robot navigation, and SODAR (an upward looking in-air sonar) is used for atmospheric investigations. The term *sonar* is also used for the equipment used to generate and receive the sound. The acoustic frequencies used in sonar systems vary from very low (infrasonic) to extremely high (ultrasonic). The study of underwater sound is known as underwater acoustics or hydroacoustics.

History

Although some animals (dolphins and bats) have used sound for communication and object detection for millions of years, use by humans in the water is initially recorded by Leonardo Da Vinci in 1490: a tube inserted into the water was said to be used to detect vessels by placing an ear to the tube.

In the 19th century an underwater bell was used as an ancillary to lighthouses to provide warning of hazards.

The use of sound to 'echo locate' underwater in the same way as bats use sound for aerial navigation seems to have been prompted by the *Titanic* disaster of 1912. The world's first patent for an underwater echo ranging device was filed at the British Patent Office by English meteorologist Lewis Richardson a month after the sinking of the Titanic, and a German physicist Alexander Behm obtained a patent for an echo sounder in 1913.

The Canadian engineer Reginald Fessenden, while working for the Submarine Signal Company in Boston, built an experimental system beginning in 1912, a system later tested in Boston Harbor, and finally in 1914 from the U.S. Revenue (now Coast Guard) Cutter Miami on the Grand Banks off Newfoundland Canada. In that test, Fessenden demonstrated depth sounding, underwater communications (Morse Code) and echo ranging (detecting an iceberg at two miles (3 km) range). The so-called Fessenden oscillator, at ca. 500 Hz frequency, was unable to determine the bearing of the berg due to the 3 metre wavelength and the small dimension of the transducer's radiating face (less than 1 metre in diameter). The ten Montreal-built British H class submarines launched in 1915 were equipped with a Fessenden oscillator.

During World War I the need to detect submarines prompted more research into the use of sound. The British made early use of underwater hydrophones, while the French physicist Paul Langevin, working with a Russian immigrant electrical engineer, Constantin Chilowski, worked on the development of active sound devices for detecting submarines in 1915 using quartz. Although piezoelectric and magnetostrictive transducers later superseded the electrostatic transducers they used, this work influenced future designs. Lightweight sound-sensitive plastic film and fibre optics have been used for hydrophones (acousto-electric transducers for in-water use), while Terfenol-D and PMN (lead magnesium niobate) have been developed for projectors.

Asdic

In 1916, under the British Board of Invention and Research, Canadian physicist Robert William Boyle took on the active sound detection project with A B Wood, producing a prototype for testing in mid 1917. This work, for the Anti-Submarine Division of the British Naval Staff, was undertaken in utmost secrecy, and used quartz piezoelectric crystals to produce the world's first practical underwater active sound detection apparatus. To maintain secrecy no mention of sound experimentation or quartz was made- the word used to describe the early work ('supersonics') was changed to 'ASD'ics, and the quartz material to 'ASD'ivite: hence the British acronym *ASDIC*. In 1939, in response to a question from the Oxford English Dictionary, the Admiralty made up the story that it stood for 'Allied Submarine Detection Investigation Committee', and this is still widely believed, though no committee bearing this name has been found in the Admiralty archives.

By 1918, both France and Britain had built prototype active systems. The British tested their ASDIC on HMS *Antrim* in 1920, and started production in 1922. The 6th Destroyer Flotilla had ASDIC-equipped vessels in 1923. An anti-submarine school, HMS *Osprey*, and a training flotilla of four vessels were established on Portland in 1924. The US Sonar QB set arrived in 1931.

By the outbreak of World War II, the Royal Navy had five sets for different surface ship classes, and others for submarines, incorporated into a complete anti-submarine attack system. The effectiveness of early ASDIC was hamstrung by the use of the depth charge as an anti-submarine weapon.

This required an attacking vessel to pass over a submerged contact before dropping charges over the stern, resulting in a loss of ASDIC contact in the moments leading up to attack. The hunter was effectively firing blind, during which time a submarine commander could take evasive action.

This situation was remedied by using several ships cooperating and by the adoption of "ahead throwing weapons", such as Hedgehog and later Squid, which projected warheads at a target ahead of the attacker and thus still in ASDIC contact. Developments during the war resulted in British ASDIC sets which used several different shapes of beam, continuously covering blind spots. Later, acoustic torpedoes were used.

At the start of World War II, British ASDIC technology was transferred for free to the United States. Research on ASDIC and underwater sound was expanded in the UK and in the US. Many new types of military sound detection were developed. These included sonobuoys, first developed by the British in 1944 under the codename

High Tea, dipping/dunking sonar and mine detection sonar. This work formed the basis for post war developments related to countering the nuclear submarine. Work on sonar had also been carried out in the Axis countries, notably in Germany, which included countermeasures. At the end of World War II this German work was assimilated by Britain and the US. Sonars have continued to be developed by many countries, including Russia, for both military and civil uses. In recent years the major military development has been the increasing interest in low frequency active systems.

Sonar

During the 1930s American engineers developed their own underwater sound detection technology and important discoveries were made, such as thermoclines, that would help future development. After technical information was exchanged between the two countries during the Second World War, Americans began to use the term *SONAR* for their systems, coined as the equivalent of RADAR.

Performance Factors

The detection, classification and localisation performance of a sonar depends on the environment and the receiving equipment, as well as the transmitting equipment in an active sonar or the target radiated noise in a passive sonar.

Sound Propagation

Sonar operation is affected by variations in sound speed, particularly in the vertical plane. Sound travels more slowly in fresh water than in sea water, though the difference is small. The speed is determined by the water's bulk modulus and mass density. The bulk modulus is affected by temperature, dissolved impurities (usually salinity), and pressure. The density effect is small. The speed of sound (in feet per second) is approximately:

4388 + (11.25 × temperature (in °F)) + (0.0182 × depth (in feet)) + salinity (in parts-per-thousand).

This empirically derived approximation equation is reasonably accurate for normal temperatures, concentrations of salinity and the range of most ocean depths. Ocean temperature varies with depth, but at between 30 and 100 meters there is often a marked change, called the thermocline, dividing the warmer surface water from the cold, still waters that make up the rest of the ocean. This can frustrate sonar, because a sound originating on one side of the thermocline tends to be bent, or refracted, through the thermocline. The thermocline may be present in shallower coastal waters. However, wave action will often mix the water column and eliminate the thermocline. Water pressure also affects sound propagation: higher pressure increases the sound speed, which causes the sound waves to refract away from the area of higher sound speed. The mathematical model of refraction is called Snell's law.

If the sound source is deep and the conditions are right, propagation may occur in the 'deep sound channel'. This provides extremely low propagation loss to a receiver in the channel. This is because of sound trapping in the channel with no losses at the boundaries. Similar propagation can occur in the 'surface duct' under suitable conditions. However in this case there are reflection losses at the surface.

In shallow water propagation is generally by repeated reflection at the surface and bottom, where considerable losses can occur.

Sound propagation is affected by absorption in the water itself as well as at the surface and bottom. This absorption depends upon frequency, with several different mechanisms in sea water. Long-range sonar uses low frequencies to minimise absorption effects.

The sea contains many sources of noise that interfere with the desired target echo or signature. The main noise sources are waves and shipping. The motion of the receiver through the water can also cause speed-dependent low frequency noise.

Scattering

When active sonar is used, scattering occurs from small objects in the sea as well as from the bottom and surface. This can be a major source of interference. This acoustic scattering is analogous to the scattering of the light from a car's headlights in fog: a high-intensity pencil beam will penetrate the fog to some extent, but broader-beam headlights emit much light in unwanted directions, much of which is scattered back to the observer, overwhelming that reflected from the target ("white-out"). For analogous reasons active sonar needs to transmit in a narrow beam to minimise scattering.

Target Characteristics

The sound *reflection* characteristics of the target of an active sonar, such as a submarine, are known as its target strength. A complication is that echoes are also obtained from other objects in the sea such as whales, wakes, schools of fish and rocks.

Passive sonar detects the target's *radiated* noise characteristics. The radiated spectrum comprises a continuous spectrum of noise with peaks at certain frequencies which can be used for classification.

Countermeasures

Active (powered) countermeasures may be launched by a submarine under attack to raise the noise level, provide a large false target, and obscure the signature of the submarine itself.

Passive (i.e., non-powered) countermeasures include:

- Mounting noise-generating devices on isolating devices.
- Sound-absorbent coatings on the hulls of submarines, for example anechoic tiles.

Active Sonar

Active sonar uses a sound transmitter and a receiver. When the two are in the same place it is monostatic operation. When the transmitter and receiver are separated it is bistatic operation. When more transmitters (or more receivers) are used, again spatially separated, it is multistatic operation. Most sonars are used monostatically with the same array often being used for transmission and reception. Active sonobuoy fields may be operated multistatically.

Active sonar creates a pulse of sound, often called a "ping", and then listens for reflections (echo) of the pulse. This pulse of sound is generally created electronically using a sonar Projector consisting of a signal generator, power amplifier and electro-acoustic transducer/array. A beamformer is usually employed to concentrate the acoustic power into a beam, which may be swept to cover the required search angles. Generally, the electro-acoustic transducers are of the Tonpilz type and their design may be optimised to achieve maximum efficiency over the widest bandwidth, in order to optimise performance of the overall system. Occasionally, the acoustic pulse may be created by other means, e.g. (1) chemically using explosives, or (2) airguns or (3) plasma sound sources.

To measure the distance to an object, the time from transmission of a pulse to reception is measured and converted into a range by knowing the speed of sound. To measure the bearing, several hydrophones are used, and the set measures the relative arrival time to each, or with an array of hydrophones, by measuring the relative amplitude in beams formed through a process called beamforming. Use of an array reduces the spatial response so that to provide wide cover multibeam systems are used. The target signal (if present) together with noise is then passed through various forms of signal processing, which for simple sonars may be just energy measurement. It is then presented to some form of decision device that calls the output either the required signal or noise. This decision device may be an operator with headphones or a display, or in more sophisticated sonars this function may be carried out by software. Further processes may be carried out to classify the target and localise it, as well as measuring its velocity.

The pulse may be at constant frequency or a chirp of changing frequency (to allow pulse compression on reception). Simple sonars generally use the former with a filter wide enough to cover possible Doppler changes due to target movement, while more complex ones generally include the latter technique. Since digital processing became available pulse compression has usually been implemented using digital correlation techniques. Military sonars often have multiple beams to provide all-round cover while simple ones only cover a narrow arc, although the beam may be rotated, relatively slowly, by mechanical scanning.

Particularly when single frequency transmissions are used, the Doppler effect can be used to measure the radial speed of a target. The difference in frequency between the transmitted and received signal is measured and converted into a velocity. Since Doppler shifts can be introduced by either receiver or target motion, allowance has to be made for the radial speed of the searching platform.

One useful small sonar is similar in appearance to a waterproof flashlight. The head is pointed into the water, a button is pressed, and the device displays the distance to the target. Another variant is a "fishfinder" that shows a small display with shoals of fish. Some civilian sonars (which are not designed for stealth) approach active military sonars in capability, with quite exotic three-dimensional displays of the area near the boat.

When active sonar is used to measure the distance from the transducer to the bottom, it is known as echo sounding. Similar methods may be used looking upward for wave measurement.

Active sonar is also used to measure distance through water between two sonar transducers or a combination of a hydrophone (underwater acoustic microphone) and projector (underwater acoustic speaker). A transducer is a device that can transmit and receive acoustic signals ("pings"). When a hydrophone/transducer receives a specific interrogation signal it responds by transmitting a specific reply signal. To measure distance, one transducer/projector transmits an interrogation signal and measures the time between this transmission and the receipt of the other transducer/hydrophone reply. The time difference, scaled by the speed of sound through water and divided by two, is the distance between the two platforms. This technique, when used with multiple transducers/hydrophones/ projectors, can calculate the relative positions of static and moving objects in water.

In combat situations, an active pulse can be detected by an opponent and will reveal a submarine's position.

A very directional, but low-efficiency, type of sonar (used by fisheries, military, and for port security) makes use of a complex nonlinear feature of water known as non-linear sonar, the virtual transducer being known as a *parametric array*.

Project Artemis

Project Artemis was a one-of-a-kind low-frequency sonar for surveillance that was deployed off Bermuda for several years in the early 1960s. The active portion was deployed from a World War II tanker, and the receiving array was a built into a fixed position on an offshore bank.

Transponder

This is an active sonar device that receives a stimulus and immediately (or with a delay) retransmits the received signal or a predetermined one.

Performance Prediction

A sonar target is small relative to the sphere, centred around the emitter, on which it is located. Therefore, the power of the reflected signal is very low, several orders of magnitude less than the original signal. Even if the reflected signal was of the same power, the following example (using hypothetical values) shows the problem: Suppose a sonar system is capable of emitting a 10,000 W/m^2 signal at 1 m, and detecting a 0.001 W/m^2 signal. At 100 m the signal will be 1 W/m^2 (due to the inverse-square law). If the entire signal is reflected from a 10 m^2 target, it will be at 0.001 W/m^2 when it reaches the emitter, i.e. just detectable.

However, the original signal will remain above 0.001 W/m^2 until 300 m. Any 10 m^2 target between 100 and 300 m using a similar or better system would be able to detect the pulse but would not be detected by the emitter. The detectors must be very sensitive to pick up the echoes. Since the original signal is much more powerful, it can be detected many times further than twice the range of the sonar (as in the example).

In active sonar there are two performance limitations, due to noise and reverberation. In general one or other of these will dominate so that the two effects can be initially considered separately.

In noise limited conditions at initial detection:

$$SL - 2TL + TS - (NL - DI) = DT$$

Where SL is the source level, TL is the transmission loss (or propagation loss), TS is the target strength, NL is the noise level, DI is the directivity index of the array (an approximation to the array gain) and DT is the detection threshold. In reverberation limited conditions at initial detection (neglecting array gain):

$$SL - 2TL + TS = RL + DT$$

Where RL is the reverberation level and the other factors are as before.

Marine Mammals

Active sonar may harm marine animals, although the precise mechanisms for this are not well understood. Some marine animals, such as whales and dolphins, use echolocation systems, sometimes called *biosonar* to locate predators and prey. It is conjectured that active sonar transmitters could confuse these animals and interfere with basic biological functions such as feeding and mating.

Hand-held sonar for use by a diver :

- The LIMIS (= Limpet Mine Imaging Sonar) is a hand-held or ROV-mounted imaging sonar for use by a diver. Its name is because it was designed for patrol divers (combat frogmen or Clearance Divers) to look for limpet mines in low visibility water. Links:

  - o Abstract of article by the International Society for Optical Engineering
  - o Used to find debris from the Space Shuttle Columbia crash
  - o Used in fish passage research at hydropower facilities
- The LUIS (= Lensing Underwater Imaging System) is another imaging sonar for use by a diver. Links:
  - o Used for counting salmon in a river
- There is or was a small flashlight-shaped handheld sonar for divers, that merely displays range.
- For the INSS = Integrated Navigation Sonar System see:
  - o an image.
  - o short description
  - o description.

Passive Sonar

Passive sonar listens without transmitting. It is often employed in military settings, although it is also used in science applications, *e.g.*, detecting fish for presence/absence studies in various aquatic environments- see also passive acoustics and passive radar. In the very broadest usage, this term can encompass virtually any analytical technique involving remotely generated sound, though it is usually restricted to techniques applied in an aquatic environment.

Identifying Sound Sources

Passive sonar has a wide variety of techniques for identifying the source of a detected sound. For example, U.S. vessels usually operate 60 Hz alternating current power systems. If transformers or generators are mounted without proper vibration insulation from the hull or become flooded, the 60 Hz sound from the windings can be emitted from the submarine or ship. This can help to identify its nationality, as most European submarines have 50 Hz power systems. Intermittent sound sources (such as a wrench being dropped) may also be detectable to passive sonar. Until fairly recently, an experienced trained operator identified signals, but now computers may do this.

Passive sonar systems may have large sonic databases, but the sonar operator usually finally classifies the signals manually. A computer system frequently uses these databases to identify classes of ships, actions (i.e. the speed of a ship, or the type of weapon released), and even particular ships. Publications for classification of sounds are provided by and continually updated by the US Office of Naval Intelligence.

Noise Limitations

Passive sonar on vehicles is usually severely limited because of noise generated by the vehicle. For this reason, many submarines operate nuclear reactors that can be cooled without pumps, using silent convection, or fuel cells or batteries, which can also run silently. Vehicles' propellers are also designed and precisely machined to emit minimal noise. High-speed propellers often create tiny bubbles in the water, and this cavitation has a distinct sound.

The sonar hydrophones may be towed behind the ship or submarine in order to reduce the effect of noise generated by the watercraft itself. Towed units also combat the thermocline, as the unit may be towed above or below the thermocline.

The display of most passive sonars used to be a two-dimensional waterfall display. The horizontal direction of the display is bearing. The vertical is frequency, or sometimes time. Another display technique is to colour-code frequency-time information for bearing. More recent displays are generated by the computers, and mimic radar-type plan position indicator displays.

Performance Prediction

Unlike active sonar, only one way propagation is involved. Because of the different signal processing used, the minimum detectable signal to noise ratio will be different. The equation for determining the performance of a passive sonar is:

$$SL - TL = NL - DI + DT$$

where SL is the source level, TL is the transmission loss, NL is the noise level, DI is the directivity index of the array (an approximation to the array gain) and DT is the detection threshold. The figure of merit of a passive sonar is:

$$FOM = SL + DI - (NL + DT).$$

Warfare

Modern naval warfare makes extensive use of both passive and active sonar from water-borne vessels, aircraft and fixed installations. The relative usefulness of active versus passive sonar depends on the radiated noise characteristics of the target, generally a submarine. Although in WW II active sonar was used by surface craft—submarines avoided emitting pings which revealed their presence and position—with the advent of modern signal-processing passive sonar became preferred for initial detection. Submarines were then designed for quieter operation, and active sonar is now more used. In 1987 a division of Japanese company Toshiba reportedly sold machinery to the Soviet Union that allowed it to mill submarine propeller blades so that they became radically quieter, creating a huge security issue with their newer generation of submarines.

Active sonar gives the exact bearing to a target, and sometimes the range. Active sonar works the same way as radar: a signal is emitted. The sound wave then travels in many directions from the emitting object. When it hits an object, the sound wave is then reflected in many other directions. Some of the energy will travel back to the emitting source. The echo will enable the sonar system or technician to calculate, with many factors such as the frequency, the energy of the received signal, the depth, the water temperature, the position of the reflecting object, etc. Active sonar is used when the platform commander determines that it is more important to determine the position of a possible threat submarine than it is to conceal his own position. With surface ships it might be assumed that the threat is already tracking the ship with satellite data. Any vessel around the emitting sonar will detect the emission. Having heard the signal, it is easy to identify the sonar equipment used (usually with its frequency) and its position (with the sound wave's energy). Active sonar is similar to radar in that, while it allows detection of targets at a certain range, it also enables the emitter to be detected at a far greater range, which is undesirable.

Since active sonar reveals the presence and position of the operator, and does not allow exact classification of targets, it is used by fast (planes, helicopters) and by noisy platforms (most surface ships) but rarely by submarines. When active sonar is used by surface ships or submarines, it is typically activated very briefly at intermittent periods to minimise the risk of detection. Consequently active sonar is normally considered a backup to passive sonar. In aircraft, active sonar is used in the form of disposable sonobuoys that are dropped in the aircraft's patrol area or in the vicinity of possible enemy sonar contacts. Passive sonar has several advantages. Most importantly, it is silent. If the target radiated noise level is high enough, it can have a greater range than active sonar, and allows the target to be identified. Since any motorized object makes some noise, it may in principle be detected, depending on the level of noise emitted and the ambient noise level in the area, as well as the technology used. To simplify, passive sonar "sees" around the ship using it. On a submarine, nose-mounted passive sonar detects in directions of about 270°, centred on the ship's alignment, the hull-mounted array of about 160° on each side, and the towed array of a full 360°. The invisible areas are due to the ship's own interference. Once a signal is detected in a certain direction (which means that something makes sound in that direction, this is called broadband detection) it is possible to zoom in and analyse the signal received (narrowband analysis). This is generally done using a Fourier transform to show the different frequencies making up the sound. Since every engine makes a specific sound, it is straightforward to identify the object. Databases of unique engine sounds are part of what is known as *acoustic intelligence* or ACINT.

Another use of passive sonar is to determine the target's trajectory. This process is called Target Motion Analysis (TMA), and the resultant "solution" is the target's range, course, and speed. TMA is done by marking from which direction the sound comes at different times, and comparing the motion with that of the operator's own ship. Changes in relative motion are analysed using standard geometrical techniques along with some assumptions about limiting cases.

Passive sonar is stealthy and very useful. However, it requires high-tech electronic components and is costly. It is generally deployed on expensive ships in the form of arrays to enhance detection. Surface ships use it to good effect; it is even better used by submarines, and it is also used by airplanes and helicopters, mostly to a "surprise effect", since submarines can hide under thermal layers. If a submarine's commander believes he is alone, he may bring his boat closer to the surface and be easier to detect, or go deeper and faster, and thus make more sound. Examples of sonar applications in military use are given below. Many of the civil uses given in the following section may also be applicable to naval use.

Anti-submarine Warfare

Until recently, ship sonars were usually with hull mounted arrays, either amidships or at the bow. It was soon found after their initial use that a means of reducing flow noise was required. The first were made of canvas on a framework, then steel ones were used. Now domes are usually made of reinforced plastic or pressurised rubber. Such sonars are primarily active in operation. An example of a conventional hull mounted sonar is the SQS-56.

Because of the problems of ship noise, towed sonars are also used. These also have the advantage of being able to be placed deeper in the water.

However, there are limitations on their use in shallow water. These are called towed arrays (linear) or variable depth sonars (VDS) with 2/3D arrays. A problem is that the winches required to deploy/recover these are large and expensive. VDS sets are primarily active in operation while towed arrays are passive.

An example of a modern active/passive ship towed sonar is Sonar 2087 made by Thales Underwater Systems.

Torpedoes

Modern torpedoes are generally fitted with an active/passive sonar. This may be used to home directly on the target, but wake following torpedoes are also used. An early example of an acoustic homer was the Mark 37 torpedo.

Torpedo countermeasures can be towed or free. An early example was the German Sieglinde device while the Pillenwerfer was a chemical device. A widely used US device was the towed Nixie while MOSS submarine simulator was a free device. A modern alternative to the Nixie system is the UK Royal Navy S2170 Surface Ship Torpedo Defence system.

Mines

Mines may be fitted with a sonar to detect, localize and recognize the required target. Further information is given in acoustic mine and an example is the CAPTOR mine.

Mine Countermeasures

Mine Countermeasure (MCM) Sonar, sometimes called "Mine and Obstacle Avoidance Sonar (MOAS)", is a specialised type of sonar used for detecting small objects. Most MCM sonars are hull mounted but a few types are VDS design. An example of a hull mounted MCM sonar is the Type 2193 while the SQQ-32 Mine-hunting sonar and Type 2093 systems are VDS designs.

Submarine Navigation

Submarine navigation underwater requires special skills and technologies not needed by surface ships. The challenges of underwater navigation have become more important as submarines spend more time underwater, travelling greater distances and at higher speed.

Military submarines travel underwater in an environment of total darkness with neither windows nor lights. Operating in stealth mode, they cannot use their active sonar systems to ping ahead for underwater hazards such as undersea mountains, drilling rigs or other submarines. Surfacing to obtain navigational fixes is precluded by pervasive anti-submarine warfare detection systems such as radar and satellite surveillance.

Antenna masts and antenna-equipped periscopes can be raised to obtain navigational signals but in areas of heavy surveillance, only for a few seconds or minutes; current radar technology can detect even a slender periscope while submarine shadows may be plainly visible from the air. Surfaced submarines entering and leaving port navigate similarly to traditional ships but with a few extra considerations because most of the ship rides below the waterline, making them hard for other ships to see and identify.

Navigational Technologies

Surface and Near-surface Navigation

On the surface or at periscope depth, submarines have used these methods to fix their position:

- Satellite navigation:
  - o Global positioning system (GPS)- - by entering waypoints internally, able to navigate at a more precise level.
  - o NAVSAT
- Terrestrial radio-based navigation systems; largely superseded by satellite systems
  - o LORAN — seldom if rarely used anymore.
  - o CHAYKA, the Russian counterpart of LORAN
  - o OMEGA, the Western counterpart of the Alpha Navigation System, no longer in use
  - o Alpha, the Russian counterpart of the Omega Navigation System

- Celestial navigation using the periscope, or sextant—seldom used anymore due to advancement in technology
- Radar navigation; radar signals are easily detected so radar is normally only used in friendly waters entering and exiting ports. With the implementation a more advanced radar system, many new techniques have been implemented in this process.
- Active sonar; like radar, active sonar systems are readily detected, so active sonar is usually used only entering and exiting ports.
- Pilotage — in coastal and internal waters, surfaced submarines rely on the standard system of navigational aids (buoys, navigational markers, lighthouses, etc.), utilizing the periscopes for obtaining lines of position to plot a triangulation fix.
- Voyage Management System — referred to as the VMS, utilizes digital charts with other external sources fed in, to establish the ship's position. Other information may also be entered in manually in establishing a high quality fix or position.

Deep Water Navigation

At depths below periscope depth submarines determine their position using:

- Dead reckoning course information obtained from the ship's gyrocompass, measured speed and estimates of local ocean currents, this could also be considered an estimated position as long as the ocean current is computed in.
- Inertial navigation system — this is an estimated position source, utilizes acceleration, deceleration, and pitch and roll in computing.
- Bottom contour navigation may be used in areas where detailed hydrographic data has been charted and there is adequate variation in sea floor topography. Fathometer depth measurements are compared to charted depth patterns.

Aircraft

Helicopters can be used for antisubmarine warfare by deploying fields of active/passive sonobuoys or can operate dipping sonar, such as the AQS-13. Fixed wing aircraft can also deploy sonobuoys and have greater endurance and capacity to deploy them. Processing from the sonobuoys or dipping sonar can be on the aircraft or on ship.

Helicopters have also been used for mine countermeasure missions using towed sonars such as the AQS-20A.

Underwater Communications

Dedicated sonars can be fitted to ships and submarines for underwater communication.

Ocean Surveillance

For many years, the United States operated a large set of passive sonar arrays at various points in the world's oceans, collectively called Sound Surveillance System (SOSUS) and later Integrated Undersea Surveillance System (IUSS). A similar system is believed to have been operated by the Soviet Union.

As permanently mounted arrays in the deep ocean were utilised, they were in very quiet conditions so long ranges could be achieved. Signal processing was carried out using powerful computers ashore. With the ending of the Cold War a SOSUS array has been turned over to scientific use.

In the United States Navy, a special badge known as the Integrated Undersea Surveillance System Badge is awarded to those who have been trained and qualified in its operation.

Underwater Security

Sonar can be used to detect frogmen and other scuba divers. This can be applicable around ships or at entrances to ports. Active sonar can also be used as a deterrent and/or disablement mechanism. One such device is the Cerberus system.

Hand-held Sonar

Limpet Mine Imaging Sonar (LIMIS) is a hand-held or ROV-mounted imaging sonar designed for patrol divers (combat frogmen or clearance divers) to look for limpet mines in low visibility water.

The LUIS is another imaging sonar for use by a diver.

Integrated Navigation Sonar System (INSS) is a small flashlight-shaped handheld sonar for divers that displays range.

Intercept Sonar

This is a sonar designed to detect and locate the transmissions from hostile active sonars. An example of this is the Type 2082 fitted on the British Vanguard class submarines.

Civilian Applications

Fisheries

Fishing is an important industry that is seeing growing demand, but world catch tonnage is falling as a result of serious resource problems. The industry faces a future of continuing worldwide consolidation until a point of sustainability can be reached. However, the consolidation of the fishing fleets are driving increased demands for sophisticated fish finding electronics such as sensors, sounders and sonars. Historically, fishermen have used many different techniques to find and harvest fish. However, acoustic technology has been one of the most important driving forces behind the development of the modern commercial fisheries.

Sound waves travel differently through fish than through water because a fish's air-filled swim bladder has a different density than seawater. This density difference allows the detection of schools of fish by using reflected sound. Acoustic technology is especially well suited for underwater applications since sound travels farther and faster underwater than in air. Today, commercial fishing vessels rely almost completely on acoustic sonar and sounders to detect fish. Fishermen also use active sonar and echo sounder technology to determine water depth, bottom contour, and bottom composition.

Companies such as Raymarine UK, Marport Canada, Wesmar, Furuno, Krupp, and Simrad make a variety of sonar and acoustic instruments for the deep sea commercial fishing industry. For example, net sensors take various underwater measurements and transmit the information back to a receiver onboard a vessel. Each sensor is equipped with one or more acoustic transducers depending on its specific function. Data is transmitted from the sensors using wireless acoustic telemetry and is received by a hull mounted hydrophone. The analog signals are decoded and converted by a digital acoustic receiver into data which is transmitted to a bridge computer for graphical display on a high resolution monitor.

Echo Sounding

An echo-sounder sends an acoustic pulse directly downwards to the seabed and records the returned echo. The sound pulse is generated by a transducer that emits an acoustic pulse and then "listens" for the return signal. The time for the signal to return is recorded and converted to a depth measurement by calculating the speed of sound in water. As the speed of sound in water is around 1,500 metres per second, the time interval, measured in milliseconds, between the pulse being transmitted and the echo being received, allows bottom depth and targets to be measured.

The value of underwater acoustics to the fishing industry has led to the development of other acoustic instruments that operate in a similar fashion to echo-sounders but, because their function is slightly different from the initial model of the echo-sounder, have been given different terms.

Net Location

The net sounder is an echo sounder with a transducer mounted on the headline of the net rather than on the bottom of the vessel. Nevertheless, to accommodate the distance from the transducer to the display unit, which is much greater than in a normal echo-sounder, several refinements have to be made. Two main types are available.

The first is the cable type in which the signals are sent along a cable. In this case there has to be the provision of a cable drum on which to haul, shoot and stow the cable during the different phases of the operation. The second type is the cable less net-sounder – such as Marport's Trawl Explorer- in which the signals are sent acoustically between the net and hull mounted receiver/hydrophone on the vessel. In this case no cable drum is required but sophisticated electronics are needed at the transducer and receiver.

The display on a net sounder shows the distance of the net from the bottom (or the surface), rather than the depth of water as with the echo-sounder's hull-mounted transducer. Fixed to the headline of the net, the footrope can usually be seen which gives an indication of the net performance.

Any fish passing into the net can also be seen, allowing fine adjustments to be made to catch the most fish possible. In other fisheries, where the amount of fish in the net is important, catch sensor transducers are mounted at various positions on the cod-end of the net. As the cod-end fills up these catch sensor transducers are triggered one by one and this information is transmitted acoustically to display monitors on the bridge of the vessel. The skipper can then decide when to haul the net.

Modern versions of the net sounder, using multiple element transducers, function more like a sonar than an echo sounder and show slices of the area in front of the net and not merely the vertical view that the initial net sounders used.

The sonar is an echo-sounder with a directional capability that can show fish or other objects around the vessel.good

Ship Velocity Measurement

Sonars have been developed for measuring a ship's velocity either relative to the water or to the bottom.

ROV and UUV

Small sonars have been fitted to Remotely Operated Vehicles (ROV) and Unmanned Underwater Vehicles (UUV) to allow their operation in murky conditions. These sonars are used for looking ahead of the vehicle. The Long-Term Mine Reconnaissance System is an UUV for MCM purposes.

Vehicle Location

Sonars which act as beacons are fitted to aircraft to allow their location in the event of a crash in the sea. Short and Long Baseline sonars may be used for caring out the location, such as LBL.

Scientific Applications

Biomass Estimation

Detection of fish, and other marine and aquatic life, and estimation their individual sizes or total biomass using active sonar techniques. As the sound pulse travels through water it encounters objects that are of different density or acoustic characteristics than the surrounding medium, such as fish, that reflect sound back toward the sound source. These echoes provide information on fish size, location, abundance and Behaviour. Data is usually processed and analysed using a variety of software such as Echoview.

Wave Measurement

An upward looking echo sounder mounted on the bottom or on a platform may be used to make measurements of wave height and period. From this statistics of the surface conditions at a location can be derived.

Water Velocity Measurement

Special short range sonars have been developed to allow measurements of water velocity.

Bottom Type Assessment

Sonars have been developed that can be used to characterise the sea bottom into, for example, mud, sand, and gravel. Relatively simple sonars such as echo sounders can be promoted to seafloor classification systems via add-on modules, converting echo parameters into sediment type. Different algorithms exist, but they are all based on changes in the energy or shape of the reflected sounder pings. Advanced substrate classification analysis can be achieved using calibrated (scientific) echosounders and parametric or fuzzy-logic analysis of the acoustic data.

Bottom Topography Measurement

Side-scan sonars can be used to derive maps of the topography of an area by moving the sonar across it just above the bottom. Low frequency sonars such as GLORIA have been used for continental shelf wide surveys while high frequency sonars are used for more detailed surveys of smaller areas.

Sub-bottom Profiling

Powerful low frequency echo-sounders have been developed for providing profiles of the upper layers of the ocean bottom.

Synthetic Aperture Sonar

Various synthetic aperture sonars have been built in the laboratory and some have entered use in mine-hunting and search systems. An explanation of their operation is given in synthetic aperture sonar.

Parametric Sonar

Parametric sources use the non-linearity of water to generate the difference frequency between two high frequencies. A virtual end-fire array is formed. Such a projector has advantages of broad bandwidth, narrow beamwidth, and when fully developed and carefully measured it has no obvious sidelobes: see Parametric array. Its major disadvantage is very low efficiency of only a few percent.. P.J. Westervelt's seminal 1963 JASA paper summarizes the trends involved.

Seismometer

Seismometers are instruments that measure motions of the ground, including those of seismic waves generated by earthquakes, nuclear explosions, and other seismic sources. Records of seismic waves allow seismologists to map the interior of the Earth, and locate and measure the size of these different sources. It is often used to mean *seismometer*, though it is more applicable to the older instruments in which the measuring and recording of ground motion were combined than to modern systems, in which these functions are separated. Both types provide a continuous record of ground motion; this distinguishes them from seismoscopes, which merely indicate that motion has occurred, perhaps with some simple measure of how large it was.

Basic Principles

Inertial seismometers have:

- A weight, usually called the internal mass, that can move relative to the instrument frame, but is attached to it by a system (such as a spring) that will hold it fixed relative to the frame if there is no motion, and also damp out any motions once the motion of the frame stops.
- A means of recording the motion of the mass relative to the frame, or the force needed to keep it from moving.

Any motion of the ground moves the frame. The mass tends not to move because of its inertia, and by measuring the motion between the frame and the mass the motion of the ground can be determined, even though the mass does move.

Early seismometers used optical levers or mechanical linkages to amplify the small motions involved, recording on soot-covered paper or photographic paper.

Modern instruments use electronics. In some systems, the mass is held nearly motionless relative to the frame by an electronic negative feedback loop. The motion of the mass relative to the frame is measured, and the feedback loop applies a magnetic or electrostatic force to keep the mass nearly motionless. The voltage needed to produce this force is the output of the seismometer, which is recorded digitally. In other systems the weight is allowed to move, and its motion produces a voltage in a coil attached to the mass and moving through the magnetic field of a magnet attached to the frame. This design is often used in the geophones used in seismic surveys for oil and gas.

Professional seismic observatories usually have instruments measuring three axes: north-south, east-west, and the vertical. If only one axis can be measured, this is usually the vertical because it is less noisy and gives better records of some seismic waves. The foundation of a seismic station is critical. A professional station is sometimes mounted on bedrock. The best mountings may be in deep boreholes, which avoid thermal effects, ground noise and tilting from weather and tides. Other instruments are often mounted in insulated enclosures on small buried piers of unreinforced concrete. Reinforcing rods and aggregates would distort the pier as the temperature changes. A site is always surveyed for ground noise with a temporary installation before pouring the pier and laying conduit.

Zhang Heng's Seismoscope

In AD 132, Zhang Heng of China's Han dynasty invented the first seismoscope (by the definition above), which was called *Houfeng Didong Yi* (literally, "instrument for measuring the seasonal winds and the movements of the Earth"). The description we have, from the History of the Later Han Dynasty, says that it was a large bronze vessel, about 2 meters in diameter; at eight points around the top were dragon's heads holding bronze balls.

When there was an earthquake, one of the mouths would open and drop its ball into a bronze toad at the base, making a sound and supposedly showing the direction of the earthquake. On at least one occasion, probably at the time of a large earthquake in Gansu in 143 CE, the seismoscope indicated an earthquake even though one was not felt.

The available text says that inside the vessel was a central column that could move along eight tracks; this is thought to refer to a pendulum, though it is not known exactly how this was linked to a mechanism that would open only one dragon's mouth. The first ever earthquake recorded by this seismograph was supposedly *somewhere in the east.* Days later, a rider from the east reported this earthquake.

An Early Example

The principle can be shown by an early special purpose seismometer. This consisted of a large stationary pendulum, with a stylus on the bottom. As the earth starts to move, the heavy mass of the pendulum has the inertia to stay still in the non-earth frame of reference. The result is that the stylus scratches a pattern corresponding with the Earth's movement. This type of strong motion seismometer recorded upon a smoked glass (glass with carbon soot). While not sensitive enough to detect distant earthquakes, this instrument could indicate the direction of the pressure waves and thus help find the epicenter of a local earthquake — such instruments were useful in the analysis of the 1906 San Francisco earthquake. Further re-analysis was performed in the 1980s using these early recordings, enabling a more precise determination of the initial fault break location in Marin county and its subsequent progression, mostly to the south.

Early Designs

After 1880, most seismometers were descended from those developed by the team of John Milne, James Alfred Ewing and Thomas Gray, who worked in Japan from 1880 to 1895. These seismometers used damped horizontal pendulums. After World War II, these were adapted into the widely used Press-Ewing seismometer.

Later, professional suites of instruments for the world-wide standard seismographic network had one set of instruments tuned to oscillate at fifteen seconds, and the other at ninety seconds, each set measuring in three directions. Amateurs or observatories with limited means tuned their smaller, less sensitive instruments to ten seconds. The basic damped horizontal pendulum seismometer swings like the gate of a fence. A heavy weight is mounted on the point of a long (from 10 cm to several meters) triangle, hinged at its vertical edge. As the ground moves, the weight stays unmoving, swinging the "gate" on the hinge.

The advantage of a horizontal pendulum is that it achieves very low frequencies of oscillation in a compact instrument. The "gate" is slightly tilted, so the weight tends to slowly return to a central position. The pendulum is adjusted (before the damping is installed) to oscillate once per three seconds, or once per thirty seconds. The general-purpose instruments of small stations or amateurs usually oscillate once per ten seconds. A pan of oil is placed under the arm, and a small sheet of metal mounted on the underside of the arm drags in the oil to damp oscillations. The level of oil, position on the arm, and angle and size of sheet is adjusted until the damping is "critical," that is, almost having oscillation. The hinge is very low friction, often torsion wires, so the only friction is the internal friction of the wire. Small seismographs with low proof masses are placed in a vacuum to reduce disturbances from air currents.

Zollner described torsionally-suspended horizontal pendulums as early as 1869, but developed them for gravimetry rather than seismometry. Early seismometers had an arrangement of levers on jeweled bearings, to scratch smoked glass or paper. Later, mirrors reflected a light beam to a direct-recording plate or roll of photographic paper. Briefly, some designs returned to mechanical movements to save money. In mid-twentieth-century systems, the light was reflected to a pair of differential electronic photosensors called a photomultiplier. The voltage generated in the photomultiplier was used to drive galvanometers which had a small mirror mounted on the axis. The moving reflected light beam would strike the surface of the turning drum, which was covered with photo-sensitive paper. The expense of developing photo sensitive paper caused many seismic observatories to switch to ink or thermal-sensitive paper.

Another relatively simple device was used in the late 19th and early 20th century. This consisted of a pendulum free to swing in any direction, with a scribe at the bottom touching a smoked glass plate. While not providing time information or information on distant earthquakes these did give accurate initial shock directions and proved useful in a late 20th century analysis of the 1906 San Francisco earthquake.

Modern Instruments

Modern instruments use electronic sensors, amplifiers, and recording devices. Most are broadband covering a wide range of frequencies. Some seismometers can measure motions with frequencies from 500 Hz to 0.00118 Hz (1/500 = 0.002 seconds per cycle, to 1/ 0.00118 = 850 seconds per cycle). The mechanical suspension for horizontal instruments remains the garden-gate described above. Vertical instruments use some kind of constant-force suspension, such as the LaCoste suspension. The LaCoste suspension uses a zero-length spring to provide a long period (high sensitivity). Some modern instruments use a "triaxial" design, in which three identical motion sensors are set at the same angle to the vertical but 120 degrees apart on the horizontal. Vertical and horizontal motions can be computed from the outputs of the three sensors.

Seismometers unavoidably introduce some distortion into the signals they measure, but professionally-designed systems have carefully characterized frequency transforms.

Modern sensitivities come in three broad ranges: geophones, 50 to 750 V/m; local geologic seismographs, about 1,500 V/m; and teleseismographs, used for world survey, about 20,000 V/m. Instruments come in three main varieties: short period, long period and broadband. The short and long period measure velocity and are very sensitive, however they 'clip' the signal or go off-scale for ground motion that is strong enough to be felt by people. A 24-bit analog-to-digital conversion channel is commonplace. Practical devices are linear to roughly one part per million.

Delivered seismometers come with two styles of output: analog and digital. Analog seismographs require analog recording equipment, possibly including an analog-to-digital converter. The output of a digital seismographs can be simply input to a computer. They present the data in standard digital forms (often "SE2" over ethernet).

Teleseismometers

The modern broadband seismograph can record a very broad range of frequencies. It consists of a small "proof mass", confined by electrical forces, driven by sophisticated electronics. As the earth moves, the electronics attempt to hold the mass steady through a feedback circuit. The amount of force necessary to achieve this is then recorded. In most designs the electronics holds a mass motionless relative to the frame.

This device is called a "force balance accelerometer". It measures acceleration instead of velocity of ground movement. Basically, the distance between the mass and some part of the frame is measured very precisely, by a linear variable differential transformer. Some instruments use a linear variable differential capacitor.

That measurement is then amplified by electronic amplifiers attached to parts of an electronic negative feedback loop. One of the amplified currents from the negative feedback loop drives a coil very like a loudspeaker, except that the coil is attached to the mass, and the magnet is mounted on the frame. The result is that the mass stays nearly motionless.

Most instruments measure directly the ground motion using the distance sensor. The voltage generated in a sense coil on the mass by the magnet directly measures the instantaneous velocity of the ground. The current to the drive coil provides a sensitive, accurate measurement of the force between the mass and frame, thus measuring directly the ground's acceleration (using f=ma where f=force, m=mass, a=acceleration).

One of the continuing problems with sensitive vertical seismographs is the buoyancy of their masses. The uneven changes in pressure caused by wind blowing on an open window can easily change the density of the air in a room enough to cause a vertical seismograph to show spurious signals. Therefore, most professional seismographs are sealed in rigid gas-tight enclosures. For example, this is why a common Streckheisen model has a thick glass base that must be glued to its pier without bubbles in the glue.

It might seem logical to make the heavy magnet serve as a mass, but that subjects the seismograph to errors when the Earth's magnetic field moves. This is also why seismograph's moving parts are constructed from a material that interacts minimally with magnetic fields. A seismograph is also sensitive to changes in temperature so many instruments are constructed from low expansion materials such as nonmagnetic invar.

The hinges on a seismograph are usually patented, and by the time the patent has expired, the design has been improved. The most successful public domain designs use thin foil hinges in a clamp.

Another issue is that the transfer function of a seismograph must be accurately characterized, so that its frequency response is known. This is often the crucial difference between professional and amateur instruments. Most instruments are characterized on a variable frequency shaking table.

Strong-motion Seismometers

Another type of seismometer is a digital strong-motion seismometer, or accelerograph. The data from such an instrument is essential to understand how an earthquake affects manmade structures.

A strong-motion seismometer measures acceleration. This can be mathematically integrated later to give velocity and position. Strong-motion seismometers are not as sensitive to ground motions as teleseismic instruments but they stay on scale during the strongest seismic shaking.

Other Forms

Accelerographs and geophones are often heavy cylindrical magnets with a spring-mounted coil inside. As case moves, the coil tends to stay stationary, so the magnetic field cuts the wires, inducing current in the output wires. They receive frequencies from several hundred hertz down to 4.5 Hz or even as low as 1 Hz with higher quality models. Some have electronic damping, a low-budget way to get some of the performance of the closed-loop wide-band geologic seismographs.

Strain-beam accelerometers constructed as integrated circuits are too insensitive for geologic seismographs (2002), but are widely used in geophones.

Some other sensitive designs measure the current generated by the flow of a non-corrosive ionic fluid through an electret sponge or a conductive fluid through a magnetic field.

Modern Recording

Today, the most common recorder is a computer with an analog-to-digital converter, a disk drive and an internet connection; for amateurs, a PC with a sound card and associated software is adequate. Most systems record continuously, but some record only when a signal is detected, as shown by a short-term increase in the variation of the signal, compared to its long-term average (which can vary slowly because of changes in seismic noise).

Interconnected Seismometers

Seismometers spaced in an array can also be used to precisely locate, in three dimensions, the source of an earthquake, using the time it takes for seismic waves to propagate away from the hypocenter, the initiating point of fault rupture. Interconnected seismometers are also used to detect underground nuclear test explosions. These seismometer are often used as part of a large scale, multi-million dollar governmental or scientific project, but some organizations, such as the Quake-Catcher Network, can use residential size detectors built into computers to detect earthquakes as well.

In reflection seismology, an array of seismometers image sub-surface features. The data are reduced to images using algorithms similar to tomography. The data reduction methods resemble those of computer-aided tomographic medical imaging X-ray machines (CAT-scans), or imaging sonars. A world-wide array of seismometers can actually image the interior of the Earth in wave-speed and transmissivity. This type of system uses events such as earthquakes, impact events or nuclear explosions as wave sources.

The first efforts at this method used manual data reduction from paper seismograph charts. Modern digital seismograph records are better adapted to direct computer use. With inexpensive seismometer designs and internet access, amateurs and small institutions have even formed a "public seismograph network."

Seismographic systems used for petroleum or other mineral exploration historically used an explosive and a wireline of geophones unrolled behind a truck. Now most short-range systems use "thumpers" that hit the ground, and some small commercial systems have such good digital signal processing that a few sledgehammer strikes provide enough signal for short-distance refractive surveys.

Exotic cross or two-dimensional arrays of geophones are sometimes used to perform three-dimensional reflective imaging of subsurface features. Basic linear refractive geomapping software (once a black art) is available off-the-shelf, running on laptop computers, using strings as small as three geophones. Some systems now come in an 18" (0.5 m) plastic field case with a computer, display and printer in the cover.

Small seismic imaging systems are now sufficiently inexpensive to be used by civil engineers to survey foundation sites, locate bedrock, and find subsurface water.

To coordinate a series of large-scale observations, most sensing systems depend on the following: platform location, what time it is, and the rotation and orientation of the sensor. High-end instruments now often use positional information from satellite navigation systems.

The rotation and orientation is often provided within a degree or two with electronic compasses. Compasses can measure not just azimuth (i.e. degrees to magnetic north), but also altitude (degrees above the horizon), since the magnetic field curves into the Earth at different angles at different latitudes. More exact orientations require gyroscopic-aided orientation, periodically realigned by different methods including navigation from stars or known benchmarks.

Resolution impacts collection and is best explained with the following relationship: less resolution=less detail & larger coverage, More resolution=more detail, less coverage. The skilled management of collection results in cost-effective collection and avoid situations such as the use of multiple high resolution data which tends to clog transmission and storage infrastructure.

Data Processing

Generally speaking, remote sensing works on the principle of the *inverse problem*. While the object or phenomenon of interest (the state) may not be directly measured, there exists some other variable that can be detected and measured (the observation), which may be related to the object of interest through the use of a data-derived computer model. The common analogy given to describe this is trying to determine the type of animal from its footprints. For example, while it is impossible to directly measure temperatures in the upper atmosphere, it is possible to measure the spectral emissions from a known chemical species (such as carbon dioxide) in that region. The frequency of the emission may then be related to the temperature in that region via various thermodynamic relations. The quality of remote sensing data consists of its spatial, spectral, radiometric and temporal resolutions.

Spatial Resolution

The size of a pixel that is recorded in a raster image – typically pixels may correspond to square areas ranging in side length from 1 to 1,000 metres (3.3 to 3,300 ft).

Spectral Resolution

The wavelength width of the different frequency bands recorded – usually, this is related to the number of frequency bands recorded by the platform. Current Landsat collection is that of seven bands, including several in the infra-red spectrum, ranging from a spectral resolution of 0.07 to 2.1 μm. The Hyperion sensor on Earth Observing-1 resolves 220 bands from 0.4 to 2.5 μm, with a spectral resolution of 0.10 to 0.11 μm per band.

Radiometric Resolution

The number of different intensities of radiation the sensor is able to distinguish. Typically, this ranges from 8 to 14 bits, corresponding to 256 levels of the gray scale and up to 16,384 intensities or "shades" of colour, in each band. It also depends on the instrument noise.

Temporal Resolution

The frequency of flyovers by the satellite or plane, and is only relevant in time-series studies or those requiring an averaged or mosaic image as in deforesting monitoring. This was first used by the intelligence community where repeated coverage revealed changes in infrastructure, the deployment of units or the modification/introduction of equipment. Cloud cover over a given area or object makes it necessary to repeat the collection of said location.

In order to create sensor-based maps, most remote sensing systems expect to extrapolate sensor data in relation to a reference point including distances between known points on the ground. This depends on the type of sensor used. For example, in conventional photographs, distances are accurate in the centre of the image, with the distortion of measurements increasing the farther you get from the centre. Another factor is that of the platen against which the film is pressed can cause severe errors when photographs are used to measure ground distances.

The step in which this problem is resolved is called georeferencing, and involves computer-aided matching up of points in the image (typically 30 or more points per image) which is extrapolated with the use of an established benchmark, "warping" the image to produce accurate spatial data. As of the early 1990s, most satellite images are sold fully georeferenced.

In addition, images may need to be radiometrically and atmospherically corrected.

Radiometric Correction

Gives a scale to the pixel values, e.g. the monochromatic scale of 0 to 255 will be converted to actual radiance values.

Atmospheric Correction

Eliminates atmospheric haze by rescaling each frequency band so that its minimum value (usually realised in water bodies) corresponds to a pixel value of 0. The digitizing of data also make possible to manipulate the data by changing gray-scale values.

Interpretation is the critical process of making sense of the data. The first application was that of aerial photographic collection which used the following process; spatial measurement through the use of a light table in both conventional single or stereographic coverage, added skills such as the use of photogrammetry, the use of photomosaics, repeat coverage, Making use of objects' known dimensions in order to detect modifications. Image Analysis is the recently developed automated computer-aided application which is in increasing use.

Object-Based Image Analysis (OBIA) is a sub-discipline of GIScience devoted to partitioning remote sensing (RS) imagery into meaningful image-objects, and assessing their characteristics through spatial, spectral and temporal scale.

Old data from remote sensing is often valuable because it may provide the only long-term data for a large extent of geography. At the same time, the data is often complex to interpret, and bulky to store.

Modern systems tend to store the data digitally, often with lossless compression.

The difficulty with this approach is that the data is fragile, the format may be archaic, and the data may be easy to falsify. One of the best systems for archiving data series is as computer-generated machine-readable ultrafiche, usually in typefonts such as OCR-B, or as digitized half-tone images.

Ultrafiches survive well in standard libraries, with lifetimes of several centuries. They can be created, copied, filed and retrieved by automated systems. They are about as compact as archival magnetic media, and yet can be read by human beings with minimal, standardized equipment.

Data Processing Levels

To facilitate the discussion of data processing in practice, several processing "levels" were first defined in 1986 by NASA as part of its Earth Observing System and steadily adopted since then, both internally at NASA (e.g.,) and elsewhere (e.g.,); these definitions are:

Level	*Description*
0	Reconstructed, unprocessed instrument and payload data at full resolution, with any and all communications artifacts (e.g., synchronization frames, communications headers, duplicate data) removed.
1a	Reconstructed, unprocessed instrument data at full resolution, time-referenced, and annotated with ancillary information, including radiometric and geometric calibration coefficients and georeferencing parameters (e.g., platform ephemeris) computed and appended but not applied to the Level 0 data (or if applied, in a manner that level 0 is fully recoverable from level 1a data).
1b	Level 1a data that have been processed to sensor units (e.g., radar backscatter cross section, brightness temperature, etc.); not all instruments have Level 1b data; level 0 data is not recoverable from level 1b data.
2	Derived geophysical variables (e.g., ocean wave height, soil moisture, ice concentration) at the same resolution and location as Level 1 source data.

3	Variables mapped on uniform spacetime grid scales, usually with some completeness and consistency (e.g., missing points interpolated, complete regions mosaicked together from multiple orbits, etc).
4	Model output or results from analyses of lower level data (i.e., variables that were not measured by the instruments but instead are derived from these measurements).

A Level 1 data record is the most fundamental (i.e., highest reversible level) data record that has significant scientific utility, and is the foundation upon which all subsequent data sets are produced. Level 2 is the first level that is directly usable for most scientific applications; its value is much greater than the lower levels. Level 2 data sets tend to be less voluminous than Level 1 data because they have been reduced temporally, spatially, or spectrally. Level 3 data sets are generally smaller than lower level data sets and thus can be dealt with without incurring a great deal of data handling overhead. These data tend to be generally more useful for many applications. The regular spatial and temporal organization of Level 3 datasets makes it feasible to readily combine data from different sources.

History

Beyond the primitive methods of remote sensing our earliest ancestors used (ex.: standing on a high cliff or tree to view the landscape), the modern discipline arose with the development of flight. The balloonist G. Tournachon (alias Nadar) made photographs of Paris from his balloon in 1858. Messenger pigeons, kites, rockets and unmanned balloons were also used for early images. With the exception of balloons, these first, individual images were not particularly useful for map making or for scientific purposes.

Systematic aerial photography was developed for military surveillance and reconnaissance purposes beginning in World War I and reaching a climax during the Cold War with the use of modified combat aircraft such as the P-51, P-38, RB-66 and the F-4C, or specifically designed collection platforms such as the U2/TR-1, SR-71, A-5 and the OV-1 series both in overhead and stand-off collection. A more recent development is that of increasingly smaller sensor pods such as those used by law enforcement and the military, in both manned and unmanned platforms. The advantage of this approach is that this requires minimal modification to a given airframe. Later imaging technologies would include Infra-red, conventional, doppler and synthetic aperture radar.

The development of artificial satellites in the latter half of the 20th century allowed remote sensing to progress to a global scale as of the end of the Cold War. Instrumentation aboard various Earth observing and weather satellites such as Landsat, the Nimbus and more recent missions such as RADARSAT and UARS provided global measurements of various data for civil, research, and military purposes. Space probes to other planets have also provided the opportunity to conduct remote sensing studies in extraterrestrial environments, synthetic aperture radar aboard the Magellan spacecraft provided detailed topographic maps of Venus, while instruments aboard SOHO allowed studies to be performed on the Sun and the solar wind, just to name a few examples.

Recent developments include, beginning in the 1960s and 1970s with the development of image processing of satellite imagery. Several research groups in Silicon Valley including NASA Ames Research Centre, GTE and ESL Inc. developed Fourier transform techniques leading to the first notable enhancement of imagery data. The introduction of online web services for easy access to remote sensing data in the 21st century (mainly low/medium-resolution images), like Google Earth, has made remote sensing more familiar to the big public and has popularized the science.

Remote Sensing Software

Remote Sensing data is processed and analysed with computer software, known as a remote sensing application. A large number of proprietary and open source applications exist to process remote sensing data. According to an NOAA Sponsored Research by Global Marketing Insights, Inc. the most used applications among Asian academic groups involved in remote sensing are as follows: ERDAS 36% (ERDAS IMAGINE 25% & ERMapper 11%); ESRI 30%; ITT Visual Information Solutions ENVI 17%; MapInfo 17%. Among Western Academic respondents as follows: ESRI 39%, ERDAS IMAGINE 27%, MapInfo 9%, AutoDesk 7%, ITT Visual Information Solutions ENVI 17%. Other important Remote Sensing Software packages include: TNTmips from MicroImages, PCI Geomatica made by PCI Geomatics, the leading remote sensing software package in Canada, IDRISI from Clark Labs, Image Analyst from Intergraph, RemoteView made by Overwatch Textron Systems, and the original object based image analysis software eCognition from Definiens. Dragon/ips is one of the oldest remote sensing packages still available, and is in some cases free. Open source remote sensing software includes GRASS GIS, QGIS, OSSIM, Opticks (software) and Orfeo toolbox.

Chapter 5

Global Positioning System

Global Positioning System (GPS) receivers can also determine altitude by trilateration with four or more satellites. In aircraft, altitude determined using autonomous GPS is not precise or accurate enough to supersede the pressure altimeter without using some method of augmentation. In hiking and climbing, it is not uncommon to find that the altitude measured by GPS is off by as much as a thousand meters, if all the available satellites happen to be close to the horizon.

Other Modes of Transport

The altimeter is an instrument optional in off-road vehicles to aid in navigation. Some high-performance luxury cars which were never intended to leave paved roads, such as the Duesenberg in the 1930s, have also been equipped with altimeters.

Hikers and mountaineers use hand-held or wrist-mounted barometric altimeters, as do skydivers.

Diesel submarines have altimeters installed on them to monitor vacuum being pulled in the event the snorkel closes while the diesels are running consequently sucking the air out of the boat.

Measuring Air Pressure

The local atmospheric pressure or *ambient pressure* is displayed in the Kollsman window of a sensitive altimeter, when it is adjusted to read zero altitude.

Satellites

A number of satellites use advanced dual-band radar altimeters to measure height from a spacecraft. That measurement, coupled with orbital elements (possibly augmented by GPS), enables determination of the terrain. The two different wavelengths of radio waves used permit the altimeter to automatically correct for varying delays in the ionosphere. Spaceborne radar altimeters have proven to be superb tools for mapping ocean-surface topography, the hills and valleys of the sea surface. These instruments send a microwave pulse to the ocean's surface and time how long it takes to return. A microwave radiometer corrects any delay that may be caused by water vapour in the atmosphere. Other corrections are also required to account for the influence of electrons in the ionosphere and the dry air mass of the atmosphere. Combining these data with the precise location of the spacecraft makes it possible to determine sea-surface height to within a few centimetres (about one inch). The strength and shape of the returning signal also provides information on wind speed and the height of ocean waves. These data are used in ocean models to calculate the speed and direction of ocean currents and the amount and location of heat stored in the ocean, which, in turn, reveals global climate variations.

LIDAR

LIDAR (Light Detection And Ranging) is an optical remote sensing technology that measures properties of scattered light to find range and/or other information of a distant target. The prevalent method to determine distance to an object or surface is to use laser pulses. Like the similar radar technology, which uses radio waves, the range to an object is determined by measuring the time delay between transmission of a pulse and detection of the reflected signal. LIDAR technology has application in Geomatics, archaeology, geography, geology, geomorphology, seismology, forestry, remote sensing and atmospheric physics. Applications of LIDAR include ALSM (*Airborne Laser Swath Mapping*), laser altimetry or LIDAR Contour Mapping. The acronym LADAR (*Laser Detection and Ranging*) is often used in military contexts. The term "laser radar" is also in use even though LIDAR does not employ microwaves or radio waves, which is definitional to radar.

General Description

The primary difference between LIDAR and radar is that LIDAR uses much shorter wavelengths of the electromagnetic spectrum, typically in the ultraviolet, visible, or near infrared range. In general it is possible to image a feature or object only about the same size as the wavelength, or larger. Thus lidar is highly sensitive to aerosols and cloud particles and has many applications in atmospheric research and meteorology.

An object needs to produce a dielectric discontinuity to reflect the transmitted wave. At radar (microwave or radio) frequencies, a metallic object produces a significant reflection. However non-metallic objects, such as rain and rocks produce weaker reflections and some materials may produce no detectable reflection at all, meaning some objects or features are effectively invisible at radar frequencies. This is especially true for very small objects (such as single molecules and aerosols).

Lidars provide one solution to these problems. The beam densities and coherency are excellent. Moreover the wavelengths are much smaller than can be achieved with radio systems, and range from about 10 micrometers to the UV (ca. 250 nm). At such wavelengths, the waves are "reflected" very well from small objects. This type of reflection is called backscattering. Different types of scattering are used for different lidar applications, most common are Rayleigh scattering, Mie scattering and Raman scattering as well as fluorescence. Based on different kinds of backscattering, the LIDAR can be accordingly called Rayleigh LiDAR, Mie LiDAR, Raman LiDAR and Na/Fe/K Fluorescence LIDAR and so on. The wavelengths are ideal for making measurements of smoke and other airborne particles (aerosols), clouds, and air molecules.

A laser typically has a very narrow beam which allows the mapping of physical features with very high resolution compared with radar. In addition, many chemical compounds interact more strongly at visible wavelengths than at microwaves, resulting in a stronger image of these materials. Suitable combinations of lasers can allow for remote mapping of atmospheric contents by looking for wavelength-dependent changes in the intensity of the returned signal.

LIDAR has been used extensively for atmospheric research and meteorology. With the deployment of the GPS in the 1980s precision positioning of aircraft became possible. GPS based surveying technology has made airborne surveying and mapping applications possible and practical. Many have been developed, using downward-looking LIDAR instruments mounted in aircraft or satellites. A recent example is the NASA Experimental Advanced Research Lidar.

Design

In general there are two kinds of lidar detection schema: "incoherent" or direct energy detection (which is principally an amplitude measurement) and Coherent detection (which is best for doppler, or phase sensitive measurements). Coherent systems generally use Optical heterodyne detection which being more sensitive than direct detection allows them to operate a much lower power but at the expense of more complex transceiver requirements.

In both coherent and incoherent LIDAR, there are two types of pulse models: micropulse lidar systems and high energy systems. Micropulse systems have developed as a result of the ever increasing amount of computer power available combined with advances in laser technology. They use considerably less energy in the laser, typically on the order of one microjoule, and are often "eye-safe," meaning they can be used without safety precautions. High-power systems are common in atmospheric research, where they are widely used for measuring many atmospheric parameters: the height, layering and densities of clouds, cloud particle properties (extinction coefficient, backscatter coefficient, depolarization), temperature, pressure, wind, humidity, trace gas concentration (ozone, methane, nitrous oxide, etc.)

There are several major components to a LIDAR system:

1. Laser — 600-1000 nm lasers are most common for non-scientific applications. They are inexpensive but since they can be focused and easily absorbed by the eye the maximum power is limited by the need to make them eye-safe. Eye-safety is often a requirement for most applications. A common alternative 1550 nm lasers are eye-safe at much higher power levels since this wavelength is not focused by the eye, but the detector technology is less advanced and so these wavelengths are generally used at longer ranges and lower accuracies. They are also used for military applications as 1550 nm is not visible in night vision goggles unlike the shorter 1000 nm infrared laser. Airborne topographic mapping lidars generally use 1064 nm diode pumped YAG lasers, while bathymetric systems generally use 532 nm frequency doubled diode pumped YAG lasers because 532 nm penetrates water with much less attenuation than does 1064 nm. Laser settings include the laser repetition rate (which controls the data collection speed).

Pulse length is generally an attribute of the laser cavity length, the number of passes required through the gain material (YAG, YLF, etc.), and Q-switch speed. Better target resolution is achieved with shorter pulses, provided the LIDAR receiver detectors and electronics have sufficient bandwidth.

2. Scanner and optics — How fast images can be developed is also affected by the speed at which it can be scanned into the system. There are several options to scan the azimuth and elevation, including dual oscillating plane mirrors, a combination with a polygon mirror, a dual axis scanner. Optic choices affect the angular resolution and range that can be detected. A hole mirror or a beam splitter are options to collect a return signal.
3. Photodetector and receiver electronics — Two main photodetector technologies are used in lidars: solid state photodetectors, such as silicon avalanche photodiodes, or photomultipliers. The sensitivity of the receiver is another parameter that has to be balanced in a LIDAR design.
4. Position and navigation systems — LIDAR sensors that are mounted on mobile platforms such as airplanes or satellites require instrumentation to determine the absolute position and orientation of the sensor. Such devices generally include a Global Positioning System receiver and an Inertial Measurement Unit (IMU).

Applications

Other than those applications listed above, there are a wide variety of applications of LIDAR, as often mentioned in National LIDAR Dataset programs.

Archaeology

LIDAR has many applications in the field of archaeology including aiding in the planning of field campaigns, mapping features beneath forest canopy, and providing an overview of broad, continuous features that may be indistinguishable on the ground. LIDAR can also provide archaeologists with the ability to create high-resolution digital elevation models (DEMs) of archaeological sites that can reveal micro-topography that are otherwise hidden by vegetation. LiDAR-derived products can be easily integrated into a Geographic Information System (GIS) for analysis and interpretation. For example at Fort Beausejour- Fort Cumberland National Historic Site, Canada, previously undiscovered archaeological features have been mapped that are related to the siege of the Fort in 1755. Features that could not be distinguished on the ground or through aerial photography were identified by overlaying hillshades of the DEM created with artificial illumination from various angles. With LiDAR the ability to produce high-resolution datasets quickly and relatively cheaply can be an advantage. Beyond efficiency, its ability to penetrate forest canopy has led to the discovery of features that were not distinguishable through traditional geo-spatial methods and are difficult to reach through field surveys.

Meteorology and Atmospheric Environment

The first LIDAR systems were used for studies of atmospheric composition, structure, clouds, and aerosols. Initially based on ruby lasers, LIDAR for meteorological applications was constructed shortly after the invention of the laser and represent one of the first applications of laser technology.

Elastic backscatter LIDAR is the simplest type of lidar and is typically used for studies of aerosols and clouds. The backscattered wavelength is identical to the transmitted wavelength, and the magnitude of the received signal at a given range depends on the backscatter coefficient of scatterers at that range and the extinction coefficients of the scatterers along the path to that range. The extinction coefficient is typically the quantity of interest.

Differential Absorption LIDAR (DIAL) is used for range-resolved measurements of a particular gas in the atmosphere, such as ozone, carbon dioxide, or water vapour. The LIDAR transmits two wavelengths: an "on-line" wavelength that is absorbed by the gas of interest and an off-line wavelength that is not absorbed. The differential absorption between the two wavelengths is a measure of the concentration of the gas as a function of range. DIAL LIDARs are essentially dual-wavelength elastic backscatter LIDARS.

Raman LIDAR is also used for measuring the concentration of atmospheric gases, but can also be used to retrieve aerosol parameters as well. Raman LIDAR exploits inelastic scattering to single out the gas of interest from all other atmospheric constituents. A small portion of the energy of the transmitted light is deposited in the gas during the scattering process, which shifts the scattered light to a longer wavelength by an amount that is unique to the species of interest. The higher the concentration of the gas, the stronger the magnitude of the backscattered signal. *Doppler LIDAR* is used to measure wind speed along the beam by measuring the frequency shift of the backscattered light. Scanning LIDARs, such as NASA's HARLIE LIDAR, have been used to measure atmospheric wind velocity in a large three dimensional cone. ESA's wind mission ADM-Aeolus will be equipped with a Doppler LIDAR system in order to provide global measurements of vertical wind profiles. A doppler LIDAR system was used in the 2008 Summer Olympics to measure wind fields during the yacht competition. Doppler LIDAR systems are also now beginning to be successfully applied in the renewable energy sector to acquire wind speed, turbulence, wind veer and wind shear data. Both pulsed and continuous wave systems are being used. Pulsed systems using signal timing to obtain vertical distance resolution, whereas continuous wave systems rely on detector focusing.

Synthetic Array LIDAR allows imaging LIDAR without the need for an array detector. It can be used for imaging Doppler velocimetry, ultra-fast frame rate (MHz) imaging, as well as for speckle reduction in coherent LIDAR.

Wind Power

Lidar is sometimes used on wind farms to more accurately measure wind speeds and wind turbulence, and an experimental lidar is mounted on a wind turbine rotor to measure oncoming horizontal winds, and proactively adjust blades to protect components and increase power.

Geology and Soil Science

High-resolution digital elevation maps generated by airborne and stationary LIDAR have led to significant advances in geomorphology, the branch of geoscience concerned with the origin and evolution of Earth's surface topography. LIDAR's abilities to detect subtle topographic features such as river terraces and river channel banks, measure the land surface elevation beneath the vegetation canopy, better resolve spatial derivatives of elevation, and detect elevation changes between repeat surveys have enabled many novel studies of the physical and chemical processes that shape landscapes. In addition to LIDAR data collected by private companies, academic consortia have been created to support the collection, processing and archiving of research-grade, publicly available LIDAR datasets. The National Centre for Airborne Laser Mapping (NCALM), supported by the National Science Foundation, collects and distributes LIDAR data in support of scientific research and education in a variety of fields, particularly geoscience and ecology.

In geophysics and tectonics, a combination of aircraft-based LIDAR and GPS have evolved into an important tool for detecting faults and measuring uplift. The output of the two technologies can produce extremely accurate elevation models for terrain that can even measure ground elevation through trees. This combination was used most famously to find the location of the Seattle Fault in Washington, USA. This combination is also being used to measure uplift at Mt. St. Helens by using data from before and after the 2004 uplift. Airborne LIDAR systems monitor glaciers and have the ability to detect subtle amounts of growth or decline. A satellite based system is NASA's ICESat which includes a LIDAR system for this purpose. NASA's Airborne Topographic Mapper is also used extensively to monitor glaciers and perform coastal change analysis. The combination is also used by soil scientists while creating a soil survey. The detailed terrain modelling allows soil scientists to see slope changes and landform breaks which indicate patterns in soil spatial relationships.

Agriculture

LIDAR also can be used to help farmers determine which areas of their fields to apply costly fertilizer. LIDAR can create a topological map of the fields and reveals the slopes and sun exposure of the farm land. Researchers at the Agricultural Research Service blended this topological information with the farm land's yield results from previous years. From this information, researchers categorized the farm land into high-, medium-, or low-yield zones. This technology is valuable to farmers because it indicates which areas to apply the expensive fertilizers to achieve the highest crop yield.

Physics and Astronomy

A worldwide network of observatories uses lidars to measure the distance to reflectors placed on the moon, allowing the moon's position to be measured with mm precision and tests of general relativity to be done. MOLA, the Mars Orbiting Laser Altimeter, used a LIDAR instrument in a Mars-orbiting satellite (the NASA Mars Global Surveyor) to produce a spectacularly precise global topographic survey of the red planet.

In September, 2008, NASA's Phoenix Lander used LIDAR to detect snow in the atmosphere of Mars.

In atmospheric physics, LIDAR is used as a remote detection instrument to measure densities of certain constituents of the middle and upper atmosphere, such as potassium, sodium, or molecular nitrogen and oxygen. These measurements can be used to calculate temperatures. LIDAR can also be used to measure wind speed and to provide information about vertical distribution of the aerosol particles.

At the JET nuclear fusion research facility, in the UK near Abingdon, Oxfordshire, LIDAR Thomson Scattering is used to determine Electron Density and Temperature profiles of the plasma.

Biology and Conservation

LIDAR has also found many applications in forestry. Canopy heights, biomass measurements, and leaf area can all be studied using airborne LIDAR systems. Similarly, LIDAR is also used by many industries, including Energy and Railroad, and the Department of Transportation as a faster way of surveying. Topographic maps can also be generated readily from LIDAR, including for recreational use such as in the production of orienteering maps.

In oceanography, LiDAR is used for estimation of phytoplankton fluorescence and generally biomass in the surface layers of the ocean. Another application is airborne lidar bathymetry of sea areas too shallow for hydrographic vessels.

In addition, the Save-the-Redwoods League is undertaking a project to map the tall redwoods on California's northern coast. LIDAR allows research scientists to not only measure the height of previously unmapped trees but to determine the biodiversity of the redwood forest. Stephen Sillett who is working with the League on the North Coast LIDAR project claims this technology will be useful in directing future efforts to preserve and protect ancient redwood trees.

Military and Law Enforcement

One situation where LIDAR has notable non-scientific application is in traffic speed enforcement, for vehicle speed measurement, as a technology alternative to radar guns. The technology for this application is small enough to be mounted in a hand held camera "gun" and permits a particular vehicle's speed to be determined from a stream of traffic. Unlike RADAR which relies on doppler shifts to directly *measure* speed, police lidar relies on the principle of time-of-flight to calculate speed. The equivalent radar based systems are often not able to isolate particular vehicles from the traffic stream. LIDAR has the distinct advantage of being able to pick out one vehicle in a cluttered traffic situation as long as the operator is aware of the limitations imposed by the range and beam divergence.

LIDAR does not suffer from "sweep" error when the operator uses the equipment correctly and when the LIDAR unit is equipped with algorithms that are able to detect when this has occurred. A combination of signal strength monitoring, receive gate timing, target position prediction and pre-filtering of the received signal wavelength prevents this from occurring. Should the beam illuminate sections of the vehicle with different reflectivity or the aspect of the vehicle changes during measurement that causes the received signal strength to be changed then the LIDAR unit will reject the measurement thereby producing speed readings of high integrity. For LIDAR units to be used in law enforcement applications a rigorous approval procedure is usually completed before deployment.

The use of many reflections and an averaging technique in the speed measurement process increase the integrity of the speed reading. Vehicles are usually equipped with a horizontally oriented registration plate that, when illuminated, causes a high integrity reflection to be returned to the LIDAR- despite the shape of the vehicle. In locations that do not require that a front or rear registration plate is fitted, headlamps and rear-reflectors provide almost ideal retro-reflective surfaces overcoming the reflections from uneven or non-compliant reflective surfaces thereby eliminating "sweep" error. It is these mechanisms which cause concern that LIDAR is somehow unreliable.

Most traffic LIDAR systems send out a stream of approximately 100 pulses over the span of three-tenths of a second. A "black box" proprietary statistical algorithm picks and chooses which progressively shorter reflections to retain from the pulses over the short fraction of a second.

Few military applications are known to be in place and are possibly classified, but a considerable amount of research is underway in their use for imaging. Higher resolution systems collect enough detail to identify targets, such as tanks. Here the name LADAR is more common. Examples of military applications of LIDAR include the Airborne Laser Mine Detection System (ALMDS) for counter-mine warfare by Arete Associates. Utilizing LIDAR and THz interferometry wide area raman spectroscopy, it is possible to detect chemical, nuclear, or biological threats at a great distance. Further investigations regarding long distance and wide area spectroscopy are currently conducted by Sandia National Laboratories.

Five LIDAR units produced by the German company Sick AG were used for short range detection on Stanley, the autonomous car that won the 2005 DARPA Grand Challenge.

A robotic Boeing AH-6 performed a fully autonomous flight in June 2010, including avoiding obstacles using LIDAR.

Vehicles

LIDAR has been used in Adaptive Cruise Control (ACC) systems for automobiles. Systems such as those by Siemens and Hella use a lidar device mounted on the front of the vehicle, such as the bumper, to monitor the distance between the vehicle and any vehicle in front of it. In the event the vehicle in front slows down or is too close, the ACC applies the brakes to slow the vehicle. When the road ahead is clear, the ACC allows the vehicle to accelerate to a speed preset by the driver.

Imaging

3-D imaging is done with both scanning and non-scanning systems. "3-D gated viewing laser radar" is a non-scanning laser radar system that applies the so-called gated viewing technique. The gated viewing technique applies a pulsed laser and a fast gated camera. There are ongoing military research programmes in Sweden, Denmark, the USA and the UK with 3-D gated viewing imaging at several kilometers range with a range resolution and accuracy better than ten centimetres.

Coherent Imaging LIDAR is possible using Synthetic array heterodyne detection which is a form of Optical heterodyne detection that enables a staring single element receiver to act as though it were an imaging array. This avoids the need for a gated camera and all ranges from all pixels are simultaneously available in the image.

Imaging LIDAR can also be performed using arrays of high speed detectors and modulation sensitive detectors arrays typically built on single chips using CMOS and hybrid CMOS/CCD fabrication techniques. In these devices each pixel performs some local processing such as demodulation or gating at high speed down converting the signals to video rate so that the array may be read like a camera. Using this technique many thousands of pixels/channels may be acquired simultaneously. In practical systems the limitation is light budget rather than parallel acquisition. LIDAR has been used in the recording of a music video without cameras. The video for the song "House of Cards" by Radiohead is believed to be the first use of real-time 3D laser scanning to record a music video.

Aerial Surveying- 3D mapping

Airborne LIDAR sensors are used by companies in the Remote Sensing field to create point clouds of the earth ground for further processing (e.g. used in forestry). A common format for saving these points (with parameters like x, y, return, intensity, elevation) is the LAS file format.

Radiometer

A radiometer is a device for measuring the radiant flux (power) of electromagnetic radiation. Generally, the term *radiometer* denotes an infrared radiation detector, yet it also includes detectors operating on any electromagnetic wavelength.

A common example is the Crookes radiometer, an early-model device wherein a rotor (with dark and light vanes) in a partial vacuum spins when exposed to light.

The Nichols radiometer operates on a different principle and is more sensitive than the Crookes type. A microwave radiometer operates in the microwave wavelengths. The radiometer contains argon gas to enable it to rotate.

The MEMS radiometer, invented by Patrick Jankowiak, can operate on the principles of Nichols or Crooke and can operate over a wide spectrum of wavelength and particle energy levels.

Photometer

In its widest sense, a photometer is an instrument for measuring light intensity or optical properties of solutions or surfaces. Photometers are used to measure:

- Illuminance
- Irradiance
- Light absorption
- Scattering of light
- Reflection of light

- Fluorescence
- Phosphorescence
- Luminescence.

History

Before electronic light sensitive elements were developed, photometry was done by estimation by the eye. The relative luminous flux of a source was compared with a standard source. The photometer is placed such that the illuminance from the source being investigated is equal to that of the standard source as equal illuminance can be judged by the eye. The relative luminous fluxes can then be calculated as the illuminance decreases proportionally to the inverse square of distance. A standard example of such a photometer consists of a piece of paper with an oil spot on it that makes the paper there slightly more transparent. When the spot is not visible from either side, the illuminance from the two sides is equal.

Principle of Photometers

Most photometers detect the light with photoresistors, photodiodes or photomultipliers. To analyse the light, the photometer may measure the light after it has passed through a filter or through a monochromator for determination at defined wavelengths or for analysis of the spectral distribution of the light.

Photon Counting

Some photometers measure light by counting individual photons rather than incoming flux. The operating principles are the same but the results are given in units such as photons/cm^2 or photons $\cdot$cm$^{-2}\cdot$sr^{-1} rather than W/cm^2 or W $\cdot$cm$^{-2}\cdot$sr^{-1}.

Due to their individual photon counting nature, these instruments are limited to observations where the irradiance is low. The irradiance is limited by the time resolution of its associated detector readout electronics. With current technology this is in the megahertz range. The maximum irradiance is also limited by the throughput and gain parameters of the detector itself.

The light sensing element in photon counting devices in NIR, visible and ultrviolet wavelengths is a photomultiplier to achieve sufficient sensitivity. In airborne and space-based remote sensing such photon counters are used at the upper reaches of the electromagnetic spectrum such as the X-ray to far ultraviolet. This is usually due to the lower radiant intensity of the objects being measured as well as the difficulty of measuring light at higher energies using its particle-like nature as compared to the wavelike nature of light at lower frequencies. Conversely, radiometers are typically used for remote sensing from the visible, infrared though radio frequency range.

Light Meter

A light meter is a device used to measure the amount of light. In photography, a light meter is often used to determine the proper exposure for a photograph. Typically a light meter will include a computer, either digital or analog, which allows the photographer to determine which shutter speed and f-number should be selected for an optimum exposure, given a certain lighting situation and film speed.

Light meters are also used in the fields of cinematography and scenic design, in order to determine the optimum light level for a scene. They are used in the general field of lighting, where they can help to reduce the amount of waste light used in the home, light pollution outdoors, and plant growing to ensure proper light levels.

Use in Photography

The earliest type of light meters were called *extinction meters* and contained a numbered or lettered row of neutral density filters of increasing density. The photographer would position the meter in front of their subject and note the filter with the greatest density that still allowed incident light to pass through. The letter or number corresponding to the filter was used as an index into a chart of appropriate aperture and shutter speed combinations for a given film speed.

Extinction meters suffered from the problem that they depended on the light sensitivity of the human eye (which can vary from person to person) and subjective interpretation.

Later meters removed the human element and relied on technologies incorporating selenium, CdS, and silicon photodetectors.

Selenium and silicon light meters use sensors that are photovoltaic: they *generate* a voltage proportional to light exposure. Selenium sensors generate enough voltage for direct connection to a meter; they need no battery to operate and this made them very convenient in completely mechanical cameras. Selenium sensors however cannot measure low light accurately (ordinary lightbulbs can take them close to their limits) and are altogether unable to measure very low light, such as candlelight, moonlight, starlight etc. Silicon sensors need an amplification circuit and require a power source such as batteries to operate. CdS light meters use a sensor based on photoresistance, i.e. their electrical resistance changes proportionately to light exposure. These also require a battery to operate. Most modern light meters use silicon or CdS sensors. They indicate the exposure either with a needle galvanometer or on an LCD screen.

Many modern consumer still and video cameras include a built-in meter that measures a scene-wide light level and are able to make an approximate measure of appropriate exposure based on that. Photographers working with controlled lighting and cinematographers use handheld light meters to precisely measure the light falling on various parts of their subjects and use suitable lighting to produce the desired exposure levels.

There are two general types of light meters: reflected-light and incident-light. Reflected-light meters measure the light *reflected by the scene* to be photographed. All in-camera meters are reflected-light meters. Reflected-light meters are calibrated to show the appropriate exposure for "average" scenes. An unusual scene with a preponderance of light colours or specular highlights would have a higher reflectance; a reflected-light meter taking a reading would incorrectly compensate for the difference in reflectance and lead to underexposure. Badly underexposed sunset photos are common exactly because of this effect: the brightness of setting sun fools the camera's light meter and, unless the in-camera logic or the photographer take care to compensate, the picture will be grossly underexposed and dull. This pitfall is avoided by incident-light meters which measure the amount of light *falling on the subject* using an integrating sphere (usually, a translucent hemispherical plastic dome is used to approximate this) placed on top of the light sensor. Because the incident-light reading is independent of the subject's reflectance, it is less likely to lead to incorrect exposures for subjects with unusual average reflectance. Taking an incident-light reading requires placing the meter at the subject's position and pointing it in the general direction of the camera, something not always achievable in practice, e.g., in landscape photography where the subject distance approaches infinity.

Another way to avoid under- or over-exposure for subjects with unusual reflectance is to use a spot meter: a reflected-light meter that measures light in a very tight cone, typically with a one degree circular angle of view. An experienced photographer can take multiple readings over the shadows, midrange and highlights of the scene to determine optimal exposure, using systems like the Zone System.

Many modern cameras include sophisticated multi-segment metering systems that measure the luminance of different parts of the scene to determine the optimal exposure. When using a film whose spectral sensitivity is not a good match to that of the light meter, for example orthochromatic black-and-white or infrared film, the meter may require special filters and re-calibration to match the sensitivity of the film.

There are other types of specialized photographic light meters. Flash meters are used in flash photography to verify correct exposure. Colour meters are used where high fidelity in colour reproduction is required. Densitometers are used in photographic reproduction.

Exposure Meter Calibration

In most cases, an incident-light meter will cause a medium tone to be recorded as a medium tone, and a reflected-light meter will cause *whatever is metered* to be recorded as a medium tone. What constitutes a "medium tone" depends on meter calibration and several other factors, including film processing or digital image conversion.

Meter calibration establishes the relationship between subject lighting and recommended camera settings. The calibration of photographic light meters is covered by ISO 2720:1974.

Exposure Equations

For reflected-light meters, camera settings are related to ISO speed and subject luminance by the reflected-light exposure equation:

$$\frac{N^2}{t} = \frac{LS}{K}$$

Where

- N is the relative aperture (f-number)
- t is the exposure time ("shutter speed") in seconds
- L is the average scene luminance

- S is the ISO arithmetic speed
- K is the reflected-light meter calibration constant.

For incident-light meters, camera settings are related to ISO speed and subject illuminance by the incident-light exposure equation:

$$\frac{N^2}{t} = \frac{ES}{C}$$

where

- E is the illuminance
- C is the incident-light meter calibration constant.

Calibration Constants

Determination of calibration constants has been largely subjective; ISO 2720:1974 states that, the constants K and C shall be chosen by statistical analysis of the results of a large number of tests carried out to determine the acceptability to a large number of observers, of a number of photographs, for which the exposure was known, obtained under various conditions of subject manner and over a range of luminances.

In practice, the variation of the calibration constants among manufacturers is considerably less than this statement might imply, and values have changed little since the early 1970s.

ISO 2720:1974 recommends a range for K of 10.6 to 13.4 with luminance in cd/m^2. Two values for K are in common use: 12.5 (Canon, Nikon, and Sekonic) and 14 (Kenko and Pentax); the difference between the two values is approximately 1/6 EV. The earliest calibration standards were developed for use with wide-angle averaging reflected-light meters. Although wide-angle average metering has largely given way to other metering sensitivity patterns (e.g., spot, centre-weighted, and multi-segment), the values for K determined for wide-angle averaging meters have remained.

The incident-light calibration constant depends on the type of light receptor. Two receptor types are common: flat (cosine-responding) and hemispherical (cardioid-responding). With a flat receptor, ISO 2720:1974 recommends a range for C of 240 to 400 with illuminance in lux; a value of 250 is commonly used. A flat receptor typically is used for measurement of lighting ratios, for measurement of illuminance, and occasionally, for determining exposure for a flat subject. For determining practical photographic exposure, a hemispherical receptor has proven more effective. Don Norwood, inventor of incident-light exposure meter with a hemispherical receptor, thought that a sphere was a reasonable representation of a photographic subject.

According to his patent (Norwood 1938), the objective was to provide an exposure meter which is substantially uniformly responsive to light incident upon the photographic subject from practically all directions which would result in the reflection of light to the camera or other photographic register and the meter provided for "measurement of the effective illumination obtaining at the position of the subject."

With a hemispherical receptor, ISO 2720:1974 recommends a range for C of 320 to 540 with illuminance in lux; in practice, values typically are between 320 (Minolta) and 340 (Sekonic). The relative responses of flat and hemispherical receptors depend upon the number and type of light sources; when each receptor is pointed at a small light source, a hemispherical receptor with $C = 330$ will indicate an exposure approximately 0.40 step greater than that indicated by a flat receptor with $C = 250$. With a slightly revised definition of illuminance, measurements with a hemispherical receptor indicate "effective scene illuminance."

Cameras with Internal Meters

Calibration of cameras with internal meters is covered by ISO 2721:1982; nonetheless, many manufacturers specify (though seldom state) exposure calibration in terms of K, and many calibration instruments (e.g., Kyoritsu-Arrowin multi-function camera testers) use the specified K to set the test parameters.

Exposure Determination with a Neutral Test Card

If a scene differs considerably from a statistically average scene, a wide-angle averaging reflected-light measurement may not indicate the correct exposure. To simulate an average scene, a substitute measurement sometimes is made of a neutral test card, or *gray card*. At best, a flat card is an approximation to a three-dimensional scene, and measurement of a test card may lead to underexposure unless adjustment is made. The instructions for a Kodak neutral test card recommend that the indicated exposure be increased by ½ step for a frontlighted scene in sunlight.

The instructions also recommend that the test card be held vertically and faced in a direction midway between the Sun and the camera; similar directions are also given in the *Kodak Professional Photoguide*. The combination of exposure increase and the card orientation gives recommended exposures that are reasonably close to those given by an incident-light meter with a hemispherical receptor when metering with an off-axis light source. In practice, additional complications may arise. Many neutral test cards are far from perfectly diffuse reflectors, and specular reflections can cause increased reflected-light meter readings that, if followed, would result in underexposure. It is possible that the neutral test card instructions include a correction for specular reflections.

Use in Illumination

Light meters or light detectors are also used in illumination. Their purpose is to measure the illumination level in the interior and to switch off or reduce the output level of luminaires. This can greatly reduce the energy burden of the building by significantly increasing the efficiency of its lighting system. It is therefore recommended to use light meters in lighting systems, especially in rooms where one cannot expect users to pay attention to manually switching off the lights. Examples include hallways, stairs, and big halls.

There are, however, significant obstacles to overcome in order to achieve a successful implementation of light meters in lighting systems, of which user acceptance is by far the most formidable. Unexpected or too frequent switching and too bright or too dark rooms are very annoying and disturbing for users of the rooms. Therefore, different switching algorithms have been developed:

- difference algorithm, where light switch on lower light level than they switch off, thus taking care that the difference between the light level of the 'on' state and 'off' state is not too big
- time delay algorithms:
  - o certain amount of time must pass since the last switch
  - o certain amount of time of sufficient illumination.

Visible Light Reflectance Photometry

A reflectance photometer measures the reflectance of a surface as a function of wavelength. The surface is illuminated with white light, and the reflected light is measured after passing through a monochromator. This type of measurement has mainly practical applications, for instance in the paint industry to characterize the colour of a surface objectively.

Absorption Spectroscopy

Absorption spectroscopy refers to spectroscopic techniques that measure the absorption of radiation, as a function of frequency or wavelength, due to its interaction with a sample. The sample absorbs energy, i.e., photons, from the radiating field. The intensity of the absorption varies as a function of frequency, and this variation is the absorption spectrum. Absorption spectroscopy is performed across the electromagnetic spectrum. Absorption spectroscopy is employed as an analytical chemistry tool to determine the presence of a particular substance in a sample and, in many cases, to quantify the amount of the substance present. Infrared and ultraviolet-visible spectroscopy are particularly common in analytical applications. Absorption spectroscopy is also employed in studies of molecular and atomic physics, astronomical spectroscopy and remote sensing.

There are a wide range of experimental approaches to measuring absorption spectra. The most common arrangement is to direct a generated beam of radiation at a sample and detect the intensity of the radiation that passes through it. The transmitted energy can be used to calculate the absorption. The source, sample arrangement and detection technique vary significantly depending on the frequency range and the purpose of the experiment.

Absorption Spectrum

A material's absorption spectrum is the fraction of incident radiation absorbed by the material over a range of frequencies. The absorption spectrum is primarily determined by the atomic and molecular composition of the material. Radiation is more likely to be absorbed at frequencies that match the energy difference between two quantum mechanical states of the molecules. The absorption that occurs due to a transition between two states is referred to as an absorption line and a spectrum is typically composed of many lines. The frequencies where absorption lines occur, as well as their relative intensities, primarily depend on the electronic and molecular structure of the molecule. The frequencies will also, though, depend on the interactions between molecules in the sample, the crystal structure (in solids) and on several environmental factors (e.g., temperature, pressure, electromagnetic field). The lines will also have a width and shape that are primarily determined by the spectral density or the density of states of the system.

Basic Theory

Absorption lines are typically classified by the nature of the quantum mechanical change induced in the molecule or atom. Rotational lines, for instance, occur when the rotational state of a molecule is changed. Rotational lines are typically found in the microwave spectral region. Vibrational lines correspond to changes in the vibrational state of the molecule and are typically found in the infrared region. Electronic lines correspond to a change in the electronic state of an atom or molecule and are typically found in the visible and ultraviolet region. X-ray absorptions are associated with the excitation of inner shell electrons in atoms. These changes can also be combined (e.g. rotation-vibration transitions), leading to new absorption lines at the combined energy of the two changes.

The energy associated with the quantum mechanical change primarily determines the frequency of the absorption line but the frequency can be shifted by several types of interactions. Electric and magnetic fields can cause a shift. Interactions with neighboring molecules can cause shifts. For instance, absorption lines of the gas phase molecule can shift significantly when that molecule is in a liquid or solid phase and interacting more strongly with neighbouring molecules.

Absorption lines are often depicted as infinitesimally thin lines, i.e., delta functions, but observed lines always have a shape that is determined by the instrument used for the observation, the material absorbing the radiation and the physical environment of that material. It is common for lines to have the shape of a Gaussian or Lorentzian distribution. It is also common for a line to be characterized solely by its intensity and width instead of the entire shape being characterized. The integrated intensity—obtained by integrating the area under the absorption line—is proportional to the amount of the absorbing substance present. The intensity is also related to the temperature of the substance and the quantum mechanical interaction between the radiation and the absorber. This interaction is quantified by the transition moment and depends on the particular lower state the transition starts from and the upper state it is connected to. The width of absorption lines may be determined by the spectrometer used to record it. A spectrometer has an inherent limit on how narrow a line it can resolve and so the observed width may be at this limit. If the width is larger than the resolution limit, then it is primarily determined by the environment of the absorber.

A liquid or solid absorber, in which neighbouring molecules strongly interact with one another, tends to have broader absorption lines than a gas. Increasing the temperature or pressure of the absorbing material will also tend to increase the line width. It is also common for several neighbouring transitions to be close enough to one another that their lines overlap and the resulting overall line is therefore broader yet.

Relation to Transmission Spectrum

Absorption and transmission spectra represent equivalent information and one can be calculated from the other through a mathematical transformation. A transmission spectrum will have its maximum intensities at wavelengths where the absorption is weakest because more light is transmitted through the sample. An absorption spectrum will have its maximum intensities at wavelengths where the absorption is strongest.

Relation to Emission Spectrum

Emission is a process by which a substance releases energy in the form of electromagnetic radiation. Emission can occur at any frequency at which absorption can occur, and this allows the absorption lines to be determined from an emission spectrum.

The emission spectrum will typically have a quite different intensity pattern from the absorption spectrum, though, so the two are not equivalent. The absorption spectrum can be calculated from the emission spectrum using appropriate theoretical models and additional information about the quantum mechanical states of the substance.

Relation to Scattering and Reflection Spectra

The scattering and reflection spectra of a material are influenced by both its index of refraction and its absorption spectrum. In an optical context, the absorption spectrum is typically quantified by the extinction coefficient, and the extinction and index coefficients are quantitatively related through the Kramers-Kronig relation. Therefore, the absorption spectrum can be derived from a scattering or reflection spectrum. This typically requires simplifying assumptions or models, and so the derived absorption spectrum is an approximation.

Applications

Analytical Chemistry

Absorption spectroscopy is useful in chemical analysis because of its specificity and its quantitative nature. The specificity of absorption spectra allows compounds to be distinguished from one another in a mixture. For example, absorption spectroscopy is used to identify the presence of pollutants in the air, distinguishing the pollutant from the nitrogen, oxygen, water and the other expected constituents. The specificity also allows unknown samples to be identified by comparing a measured spectrum with a library of reference spectra. In many cases, it is possible to determine qualitative information about a sample even if it is not in a library. Infrared spectra, for instance, have characteristics absorption bands that indicate if carbon-hydrogen or carbon-oxygen bonds are present.

An absorption spectrum can be quantitatively related to the amount of material present using the Beer-Lambert law. Determining the absolute concentration of a compound requires knowledge of the compound's absorption coefficient. The absorption coefficient for some compounds is available from reference sources, and it can also be determined by measuring the spectrum of a calibration standard with a known concentration of the target.

Remote Sensing

One of the unique advantages of spectroscopy as an analytical technique is that measurements can be made without bringing the instrument and sample into contact. Radiation that travels between a sample and an instrument will contain the spectral information, so the measurement can be made remotely. Remote spectral sensing is valuable in many situations. For example, measurements can be made in toxic or hazardous environments without placing an operator or instrument at risk. Also, sample material does not have to be brought into contact with the instrument—preventing possible cross contamination.

Remote spectral measurements present several challenges compared to laboratory measurements. The space in between the sample of interest and the instrument may also have spectral absorptions. These absorptions can mask or confound the absorption spectrum of the sample. These background interferences may also vary over time. The source of radiation in remote measurements is often an environmental source, such as sunlight or the thermal radiation from a warm object, and this makes it necessary to distinguish spectral absorption from changes in the source spectrum.

Astronomy

Astronomical spectroscopy is a particularly significant type of remote spectral sensing. In this case, the objects and samples of interest are so distant from earth that electromagnetic radiation is the only means available to measure them. Astronomical spectra contain both absorption and emission spectral information. Absorption spectroscopy has been particularly important for understanding interstellar clouds and determining that some of them contain molecules. Absorption spectroscopy is also employed in the study of extrasolar planets. Detection of extrasolar planets by the transit method also measures their absorption spectrum and allows for the determination of the planet's atmospheric composition.

Atomic and Molecular Physics

Theoretical models, principally quantum mechanical models, allow for the absorption spectra of atoms and molecules to be related to other physical properties such as electronic structure, atomic or molecular mass, and molecular geometry. Therefore, measurements of the absorption spectrum are used to determine these other properties. Microwave spectroscopy, for example, allows for the determination of bond lengths and angles with high precision.

In addition, spectral measurements can be used to determine the accuracy of theoretical predictions. For example, the Lamb shift measured in the hydrogen atomic absorption spectrum was not expected to exist at the time it was measured. Its discovery spurred and guided the development of quantum electrodynamics, and measurements of the Lamb shift are now used to determine the fine-structure constant.

Experimental Methods

Basic Approach

The most straight-forward approach to absorption spectroscopy is to generate radiation with a source, measure a reference spectrum of that radiation with a detector and then re-measure the sample spectrum after placing the material of interest in between the source and detector. The two measured spectra can then be combined to determine the material's absorption spectrum. The sample spectrum alone is not sufficient to determine the absorption spectrum because it will be affected by the experimental conditions—the spectrum of the source, the absorption spectra of other materials in between the source and detector and the wavelength dependent characteristics of the detector. The reference spectrum will be affected in the same way, though, by these experimental conditions and therefore the combination yields the absorption spectrum of the material alone.

A wide variety of radiation sources are employed in order to cover the electromagnetic spectrum. For spectroscopy, it is generally desirable for a source to cover a broad swath of wavelengths in order to measure a broad region of the absorption spectrum. Some sources inherently emit a broad spectrum. Examples of these include globars or other black body sources in the infrared, mercury lamps in the visible and ultraviolet and x-ray tubes. One recently developed, novel source of broad spectrum radiation is synchotron radiation which covers all of these spectral regions. Other radiation sources generate a narrow spectrum but the emission wavelength can be tuned to cover a spectral range. Examples of these include klystrons in the microwave region and lasers across the infrared, visible and ultraviolet region (though not all lasers have tunable wavelengths).

The detector employed to measure the radiation power will also depend on the wavelength range of interest. Most detectors are sensitive to a fairly broad spectral range and the sensor selected will often depend more on the sensitivity and noise requirements of a given measurement. Examples of detectors common in spectroscopy include heterodyne receivers in the microwave, bolometers in the millimeter-wave and infrared, mercury cadmium telluride and other cooled semiconductor detectors in the infrared, and photodiodes and photomultiplier tubes in the visible and ultraviolet.

If both the source and the detector cover a broad spectral region, then it is also necessary to introduce a means of resolving the wavelength of the radiation in order to determine the spectrum. Often a spectrograph is used to spatially separate the wavelengths of radiation so that the power at each wavelength can be measured independently. It is also common to employ interferometry to determine resolve the spectrum—Fourier transform infrared spectroscopy is a widely used implementation of this technique.

Two other issues that must be considered in setting up an absorption spectroscopy experiment include the optics used to direct the radiation and the means of holding or containing the sample material (called a cuvette or cell). For most UV, visible, and NIR measurements the use of precision quartz cuvettes are necessary.. In both cases, it is important to select materials that have relatively little absorption of their own in the wavelength range of interest. The absorption of other materials could interfere with or mask the absorption from the sample. For instance, in several wavelength ranges it is necessary to measure the sample under vacuum or in a rare gas environment because gases in the atmosphere have interfering absorption features.

Specific Approaches

- Cavity Ring Down Spectroscopy (CRDS)
- Mössbauer spectroscopy
- Photoemission spectroscopy
- Reflectance spectroscopy
- Laser Absorption Spectrometry (LAS)
- Tunable Diode Laser Absorption Spectroscopy (TDLAS)
- X-ray absorption fine structure (XAFS)
- X-ray Absorption Near Edge Structure (XANES)
- Astronomical spectroscopy.

Infrared Spectroscopy

Infrared spectroscopy (IR spectroscopy) is the spectroscopy that deals with the infrared region of the electromagnetic spectrum, that is light with a longer wavelength and lower frequency than visible light. It covers a range of techniques, mostly based on absorption spectroscopy. As with all spectroscopic techniques, it can be used to identify and study chemicals. A common laboratory instrument that uses this technique is a Fourier transform infrared (FTIR) spectrometer.

The infrared portion of the electromagnetic spectrum is usually divided into three regions; the near-, mid- and far- infrared, named for their relation to the visible spectrum. The higher energy near-IR, approximately 14000–4000 cm^{-1} (0.8–2.5 μm wavelength) can excite overtone or harmonic vibrations.

The mid-infrared, approximately 4000–400 cm^{-1} (2.5–25 μm) may be used to study the fundamental vibrations and associated rotational-vibrational structure. The far-infrared, approximately 400–10 cm^{-1}

(25–1000 μm), lying adjacent to the microwave region, has low energy and may be used for rotational spectroscopy. The names and classifications of these subregions are conventions, and are only loosely based on the relative molecular or electromagnetic properties.

Theory

Infrared spectroscopy exploits the fact that molecules absorb specific frequencies that are characteristic of their structure. These absorptions are resonant frequencies, i.e. the frequency of the absorbed radiation matches the frequency of the bond or group that vibrates.

The energies are determined by the shape of the molecular potential energy surfaces, the masses of the atoms, and the associated vibronic coupling.

In particular, in the Born–Oppenheimer and harmonic approximations, i.e. when the molecular Hamiltonian corresponding to the electronic ground state can be approximated by a harmonic oscillator in the neighbourhood of the equilibrium molecular geometry, the resonant frequencies are determined by the normal modes corresponding to the molecular electronic ground state potential energy surface.

Nevertheless, the resonant frequencies can be in a first approach related to the strength of the bond, and the mass of the atoms at either end of it. Thus, the frequency of the vibrations can be associated with a particular bond type.

Number of Vibrational Modes

In order for a vibrational mode in a molecule to be "IR active," it must be associated with changes in the permanent dipole.

A molecule can vibrate in many ways, and each way is called a *vibrational mode*. Linear molecules have 3N – 5 degrees of vibrational modes whereas nonlinear molecules have 3N – 6 degrees of vibrational modes (also called vibrational degrees of freedom). As an example H_2O, a non-linear molecule, will have $3 \times 3 - 6 = 3$ degrees of vibrational freedom, or modes. Simple diatomic molecules have only one bond and only one vibrational band. If the molecule is symmetrical, e.g. N_2, the band is not observed in the IR spectrum, but only in the Raman spectrum. Unsymmetrical diatomic molecules, e.g. CO, absorb in the IR spectrum. More complex molecules have many bonds, and their vibrational spectra are correspondingly more complex, i.e. big molecules have many peaks in their IR spectra.

The atoms in a CH_2 group, commonly found in organic compounds, can vibrate in six different ways: symmetric and antisymmetric stretching, scissoring, rocking, wagging and twisting:

Special Effects

The simplest and most important IR bands arise from the "normal modes," the simplest distortions of the molecule. In some cases, "overtone bands" are observed. These bands arise from the absorption of a photon that leads to a doubly excited vibrational state. Such bands appear at approximately twice the energy of the normal mode. Some vibrations, so-called 'combination modes," involve more than one normal mode. The phenomenon of Fermi resonance can arise when two modes are similar in energy, Fermi resonance results in an unexpected shift in energy and intensity of the bands.

Practical IR Spectroscopy

The infrared spectrum of a sample is recorded by passing a beam of infrared light through the sample. Examination of the transmitted light reveals how much energy was absorbed at each wavelength. This can be done with a monochromatic beam, which changes in wavelength over time, or by using a Fourier transform instrument to measure all wavelengths at once. From this, a transmittance or absorbance spectrum can be produced, showing at which IR wavelengths the sample absorbs. Analysis of these absorption characteristics reveals details about the molecular structure of the sample. When the frequency of the IR is the same as the vibrational frequency of a bond, absorption occurs.

This technique works almost exclusively on samples with covalent bonds. Simple spectra are obtained from samples with few IR active bonds and high levels of purity. More complex molecular structures lead to more absorption bands and more complex spectra. The technique has been used for the characterization of very complex mixtures.

Sample Preparation

Gaseous samples require a sample cell with a long pathlength (typically 5–10 cm), to compensate for the diluteness.

Liquid samples can be sandwiched between two plates of a salt (commonly sodium chloride, or common salt, although a number of other salts such as potassium bromide or calcium fluoride are also used). The plates are transparent to the infrared light and do not introduce any lines onto the spectra.

Solid samples can be prepared in a variety of ways. One common method is to crush the sample with an oily mulling agent (usually Nujol) in a marble or agate mortar, with a pestle. A thin film of the mull is smeared onto salt plates and measured. The second method is to grind a quantity of the sample with a specially purified salt (usually potassium bromide) finely (to remove scattering effects from large crystals). This powder mixture is then pressed in a mechanical press to form a translucent pellet through which the beam of the spectrometer can pass. A third technique is the "cast film" technique, which is used mainly for polymeric materials. The sample is first dissolved in a suitable, non hygroscopic solvent. A drop of this solution is deposited on surface of KBr or NaCl cell. The solution is then evaporated to dryness and the film formed on the cell is analysed directly. Care is important to ensure that the film is not too thick otherwise light cannot pass through. This technique is suitable for qualitative analysis. The final method is to use microtomy to cut a thin (20–100 μm) film from a solid sample. This is one of the most important ways of analysing failed plastic products for example because the integrity of the solid is preserved.

It is important to note that spectra obtained from different sample preparation methods will look slightly different from each other due to differences in the samples' physical states.

Comparing to a Reference

To take the infrared spectrum of a sample, it is necessary to measure both the sample and a "reference" (or "control"). This is because each measurement is affected by not only the light-absorption properties of the sample, but also the properties of the instrument (for example, what light source is used, what detector is used, etc.). The reference measurement makes it possible to eliminate the instrument influence. Mathematically, the sample transmission spectrum is divided by the reference transmission spectrum.

The appropriate "reference" depends on the measurement and its goal. The simplest reference measurement is to simply remove the sample (replacing it by air). However, sometimes a different reference is more useful. For example, if the sample is a dilute solute dissolved in water in a beaker, then a good reference measurement might be to measure pure water in the same beaker. Then the reference measurement would cancel out not only all the instrumental properties (like what light source is used), but also the light-absorbing and light-reflecting properties of the water and beaker, and the final result would just show the properties of the solute (at least approximately).

A common way to compare to a reference is sequentially: First measure the reference, then replace the reference by the sample, then measure the sample. This technique is not perfectly reliable: If the infrared lamp is a bit brighter during the reference measurement, then a bit dimmer during the sample measurement, the measurement will be distorted. More elaborate methods, such as a "two-beam" setup, can correct for these types of effects to give very accurate results.

FTIR

Fourier transform infrared (FTIR) spectroscopy is a measurement technique that allows one to record infrared spectra. Infrared light is guided through an interferometer and then through the sample (or vice versa). A moving mirror inside the apparatus alters the distribution of infrared light that passes through the interferometer. The signal directly recorded, called an "interferogram", represents light output as a function of mirror position. A data-processing technique called Fourier transform turns this raw data into the desired result (the sample's spectrum): Light output as a function of infrared wavelength (or equivalently, wavenumber). As described above, the sample's spectrum is always compared to a reference.

There is an alternate method for taking spectra (the "dispersive" or "scanning monochromator" method), where one wavelength at a time passes through the sample. The dispersive method is more common in UV-Vis spectroscopy, but is less practical in the infrared than the FTIR method. One reason that FTIR is favored is called "Fellgett's advantage" or the "multiplex advantage": The information at all frequencies is collected simultaneously, improving both speed and signal-to-noise ratio. Another is called "Jacquinot's Throughput Advantage": A dispersive measurement requires detecting much lower light levels than an FTIR measurement. There are other advantages, as well as some disadvantages, but virtually all modern infrared spectrometers are FTIR instruments.

Absorption Bands

Uses and Applications

Infrared spectroscopy is widely used in both research and industry as a simple and reliable technique for measurement, quality control and dynamic measurement. It is also used in forensic analysis in both criminal and civil cases, enabling identification of polymer degradation for example. The instruments are now small, and can be transported, even for use in field trials. With increasing technology in computer filtering and manipulation of the results, samples in solution can now be measured accurately (water produces a broad absorbance across the range of interest, and thus renders the spectra unreadable without this computer treatment). Some instruments will also automatically tell you what substance is being measured from a store of thousands of reference spectra held in storage.

By measuring at a specific frequency over time, changes in the character or quantity of a particular bond can be measured. This is especially useful in measuring the degree of polymerization in polymer manufacture. Modern research instruments can take infrared measurements across the whole range of interest as frequently as 32 times a second. This can be done whilst simultaneous measurements are made using other techniques. This makes the observations of chemical reactions and processes quicker and more accurate.

Infrared spectroscopy has been highly successful for applications in both organic and inorganic chemistry. Infrared spectroscopy has also been successfully utilized in the field of semiconductor microelectronics: for example, infrared spectroscopy can be applied to semiconductors like silicon, gallium arsenide, gallium nitride, zinc selenide, amorphous silicon, silicon nitride, etc.

Chapter 6

Analytic Signal of Radar Processing

Time–frequency Analysis

Need for a Time–frequency Approach

In signal processing, time–frequency analysis is a body of techniques and methods used for characterizing and manipulating signals whose statistics vary in time, such as transient signals.

It is a generalization and refinement of Fourier analysis, for the case when the signal frequency characteristics are varying with time. Since many signals of interest– such as speech, music, images, and medical signals – have changing frequency characteristics, time–frequency analysis has broad scope of applications.

Whereas the technique of the Fourier transform can be extended to obtain the frequency spectrum of any slowly growing locally integrable signal this requires a complete description of the signal's Behaviour over all time.

Indeed, one can think of points in the (spectral) frequency domain as smearing together information from across the entire time domain. While mathematically elegant such a technique is not appropriate for analysing a signal with indeterminate future Behaviour.

For instance one must presuppose some degree of indeterminate future Behaviour in any telecommunications systems to achieve non-zero entropy (if you already know what the other guy will say you can't learn anything).

To harness the power of a frequency representation without the need of a complete characterization in the time domain, one first obtains a time–frequency distribution of the signal, which represents the signal in both the time and frequency domains simultaneously. In such a representation the frequency domain will only reflect the Behaviour of a temporally localized version of the signal. This enables one to talk sensibly about signals whose component frequencies vary in time. For instance rather than using tempered distributions to globally transform the following function into the frequency domain one could instead use these methods to describe it as a signal with a time varying frequency.

$$x(t) = \begin{cases} \cos(\pi t); & t < 10 \\ \cos(3\pi t); & 10 \le t < 20 \\ \cos(2\pi t); & t > 20 \end{cases}$$

Once such a representation has been generated other techniques in time–frequency analysis may then be applied to the signal in order to extract information from the signal, to separate the signal from noise or interfering signals, etc.

Time–frequency Distribution Functions

Diversity of Time–frequency Formulations

There are several different ways to formulate a valid time–frequency distribution function, resulting in several well-known time–frequency distributions, such as:

- Short-time Fourier transform (including the Gabor transform),
- Wavelet transform,
- Bilinear time–frequency distribution function (Wigner distribution function),
- Modified Wigner distribution function, Gabor–Wigner distribution function, and so on.

More information about the history and the motivation of development of time–frequency distribution can be found in the entry Time–frequency representation.

Ideal TF Distribution Function

A time–frequency distribution function ideally has the following properties:

1. High clarity to make it easier to be analysed and interpreted.
2. No cross-term to avoid confusing real components from artefacts or noise.
3. A list of desirable mathematical properties to ensure such methods benefit real-life application.
4. Lower computational complexity to ensure the time needed to represent and process a signal on a time–frequency plane allows real-time implementations.

To analyse the signals well, choosing an appropriate time–frequency distribution function is important. Which time–frequency distribution function should be used depends on the application being considered, as shown by reviewing a list of applications. The high clarity of the Wigner distribution function (WDF) obtained for some signals is due to the auto-correlation function inherent in its formulation; however, the latter also causes the cross-term problem. Therefore, if we want to analyse a single-term signal, using the WDF may be the best approach; if the signal is composed of multiple components, some other methods like the Gabor transform, Gabor-Wigner distribution or Modified B-Distribution functions may be better choices. To illustrate this, we observe that by Fourier analysis, we can't recognize the two signals $x_1(t)$ and $x_2(t)$ below.

$$x_1(t) = \begin{cases} \cos(\pi t); & t < 10 \\ \cos(3\pi t); & 10 \le t < 20 \\ \cos(2\pi t); & t > 20 \end{cases}$$

$$x_2(t) = \begin{cases} \cos(\pi t); & t < 10 \\ \cos(2\pi t); & 10 \le t < 20 \\ \cos(3\pi t); & t > 20 \end{cases}$$

Thanks to the time–frequency analysis approach, we can still solve this problem of correctly identifying the two different signals.

Signal Processing Applications

The following applications need not only the time–frequency distribution functions but also some operations to the signal. The Linear canonical transform (LCT) is really helpful. By LCTs, the shape and location on the time–frequency plane of a signal can be in the arbitrary form that we want it to be.

For example, the LCTs can shift the time–frequency distribution to any location, dilate it in the horizontal and vertical direction without changing its area on the plane, shear (or twist) it, and rotate it (Fractional Fourier transform).

This powerful operation, LCT, make it more flexible to analyse and apply the time–frequency distributions. Here we list some applications of time–frequency analysis.

Instantaneous Frequency Estimation

The definition of instantaneous frequency is the time rate of change of phase, or,

$$\frac{1}{2\pi}\frac{d}{dt}\phi(t)$$

where $\phi(t)$ is the instantaneous phase of a signal. We can know the instantaneous frequency from the time–frequency plane directly if the image is clear enough. Because the high clarity is critical, we often use WDF to analyse it.

TF Filtering and Signal Decomposition

The goal of filter design is to remove the undesired component of a signal. Conventionally, we can just filter in the time domain or in the frequency domain individually as shown below. The filtering methods mentioned above can't work well for every signal which may overlap in the time domain or in the frequency domain. By using the time–frequency distribution function, we can filter in the Euclidian time–frequency domain or in the fractional domain by employing the fractional Fourier transform. An example is shown below.

Filter design in time–frequency analysis always deals with signals composed of multiple components, so one cannot use WDF due to cross-term. The Gabor transform, Gabor-Wigner distribution function, or Cohen's class distribution function may be better choices. The concept of signal decomposition relates to the need to separate one component from the others in a signal; this can be achieved through a filtering operation which require a filter design stage. Such filtering is traditionally done in the time domain or in the frequency domain; however, this may not be possible in the case of non-stationary signals that are multicomponent as such components could overlap in both the time domain and also in the frequency domain; as a consequence, the only possible way to achieve component separation and therefore a signal decomposition is to implement a time–frequency filter.

Sampling Theory

By the Nyquist–Shannon sampling theorem, we can conclude that the minimum number of sampling points without aliasing is equivalent to the area of the time–frequency distribution of a signal. (This is actually just an approximation, because the TF area of any signal is infinite.) Below is an example before and after we combine the sampling theory with the time–frequency distribution:

It is noticeable that the number of sampling points decreases after we apply the time–frequency distribution.

When we use the WDF, there might be the cross-term problem (also called interference). On the other hand, using Gabor transform causes an improvement in the clarity and readability of the representation, therefore improving its interpretation and application to practical problems. Consequently, when the signal we tend to sample is composed of single component, we use the WDF; however, if the signal consists of more than one component, using the Gabor transform, Gabor-Wigner distribution function, or other reduced interference TFDs may achieve better results.

Other Applications

Modulation and Multiplexing

Conventionally, the operation of modulation and multiplexing concentrates in time or in frequency, separately. By taking advantage of the time–frequency distribution, we can make it more efficient to modulate and multiplex. All we have to do is to fill up the time–

As illustrated in the upper example, using the WDF is not smart since the serious cross-term problem make it difficult to multiplex and modulation.

Electromagnetic Wave Propagation

We can represent an electromagnetic wave in the form of a 2 by 1 matrix

$$\begin{bmatrix} x \\ y \end{bmatrix};$$

which is similar to the time–frequency plane. When electromagnetic wave propagates through free-space, the Fresnel diffraction occurs. We can operate with the 2 by 1 matrix

$$\begin{bmatrix} x \\ y \end{bmatrix}$$

by LCT with parameter matrix

$$\begin{bmatrix} a & b \\ c & d \end{bmatrix} = \begin{bmatrix} 1 & \lambda z \\ 0 & 1 \end{bmatrix};$$

where z is the propagation distance and λ is the wavelength. When electromagnetic wave pass through a spherical lens or be reflected by a disk, the parameter matrix should be

$$\begin{bmatrix} a & b \\ c & d \end{bmatrix} = \begin{bmatrix} 1 & 0 \\ \frac{-1}{\lambda f} & 1 \end{bmatrix};$$

and

$$\begin{bmatrix} a & b \\ c & d \end{bmatrix} = \begin{bmatrix} 1 & 0 \\ \frac{1}{\lambda R} & 1 \end{bmatrix};$$

respectively, where f is the focal length of the lens and R is the radius of the disk. These corresponding results can be obtained from

$$\begin{bmatrix} a & b \\ c & d \end{bmatrix} \begin{bmatrix} x \\ y \end{bmatrix}.$$

Optics, Acoustics, and Biomedicine

Light is a kind of electromagnetic wave, so we apply the time–frequency analysis to optics in the same way as to electromagnetic wave propagation. In the same way, a characteristic of acoustic signals is that, often, its frequency varies really severely with time. Because the acoustic signals usually contain a lot of data, it is suitable to use simpler TFDs such as the Gabor transform to analyse the acoustic signals due to the lower computational complexity. If speed is not an issue, then a detailed comparison with well defined criteria should be made before selecting a particular TFD. Another approach is to define a signal dependent TFD that is adapted to the data. In biomedicine, one can use time–frequency distribution to analyse the electromyography (EMG), Electroencephalography (EEG), Electrocardiogram (ECG) or otoacoustic emissions (OAEs).

Analytic Signal

In mathematics and signal processing, the analytic representation of a real-valued function or signal facilitates many mathematical manipulations of the signal. The basic idea is that the negative frequency components of the Fourier transform (or spectrum) of a real-valued function are superfluous, due to the Hermitian symmetry of such a spectrum. These negative frequency components can be discarded with no loss of information, providing one is willing to deal with a complex-valued function instead. That makes certain attributes of the signal more accessible and facilitate the derivation of modulation and demodulation techniques, especially single-sideband. As long as the manipulated function has no negative frequency components (that is, it is still *analytic*), the conversion from complex back to real is just a matter of discarding the imaginary part. The analytic representation is a generalization of the phasor concept: while the phasor is restricted to time-invariant amplitude, phase, and frequency, the analytic signal allows for time-variable parameters.

Extensions of the Analytic Signal to Signals of Multiple Variables

The concept of analytic signal is well-defined for signals of a single variable which typically is time. For signals of two or more variables, an analytic signal can be defined in different ways, and two approaches are presented below.

Multi-dimensional Analytic Signal based on an Ad-hoc Direction

A straightforward generalization of the analytic signal can be done for a multi-dimensional signal once it is established what is meant by *negative frequencies* for this case. This can be done by introducing a normalized vector $\hat{n}$ in the Fourier domain and label any frequency vector u as negative if $u \cdot \hat{n} < 0$. The analytic signal is then produced by removing all negative frequencies and multiply the result by 2, in accordance to the procedure described for the case of one-variable signals. However, there is no particular direction for which must be chosen unless there are some additional constraints. Therefore, the choice of is ad-hoc, or application specific.

The Monogenic Signal

The real and imaginary parts of the analytic signal correspond to the two elements of the vector-valued monogenic signal, as it is defined for one-variable signals. However, the monogenic signal can be extended to arbitrary number of variables in a straightforward manner, producing an $(n + 1)$-dimensional vector-valued function for the case of n-variable signals.

Not to be Confused with

In mathematics, particularly functions of a complex variable, an analytic function refers to a function which satisfies certain properties regarding differentiability. The concept of analytic signal should not be confused with analytic functions.

Wavelet

A wavelet is a wave-like oscillation with an amplitude that starts out at zero, increases, and then decreases back to zero. It can typically be visualized as a "brief oscillation" like one might see recorded by a seismograph or heart monitor. Generally, wavelets are purposefully crafted to have specific properties that make them useful for signal processing. Wavelets can be combined, using a "shift, multiply and sum" technique called convolution, with portions of an unknown signal to extract information from the unknown signal.

For example, a wavelet could be created to have a frequency of Middle C and a short duration of roughly a 32nd note. If this wavelet were to be convolved at periodic intervals with a signal created from the recording of a song, then the results of these convolutions would be useful for determining when the Middle C note was being played in the song.

Mathematically, the wavelet will resonate if the unknown signal contains information of similar frequency- just as a tuning fork physically resonates with sound waves of its specific tuning frequency. This concept of resonance is at the core of many practical applications of wavelet theory.

As wavelets are a mathematical tool they can be used to extract information from many different kinds of data, including- but certainly not limited to- audio signals and images. Sets of wavelets are generally needed to analyse data fully. A set of "complementary" wavelets will deconstruct data without gaps or overlap so that the deconstruction process is mathematically reversible.

Thus, sets of complementary wavelets are useful in wavelet based compression/decompression algorithms where it is desirable to recover the original information with minimal loss.

In formal terms, this representation is a wavelet series representation of a square-integrable function with respect to either a complete, orthonormal set of basis functions, or an overcomplete set or Frame of a vector space, for the Hilbert space of square integrable functions.

Name

The word *wavelet* has been used for decades in digital signal processing and exploration geophysics. The equivalent French word *ondelette* meaning "small wave" was used by Morlet and Grossmann in the early 1980s.

Wavelet Theory

Wavelet theory is applicable to several subjects. All wavelet transforms may be considered forms of time-frequency representation for continuous-time (analog) signals and so are related to harmonic analysis. Almost all practically useful discrete wavelet transforms use discrete-time filterbanks. These filter banks are called the wavelet and scaling coefficients in wavelets nomenclature. These filterbanks may contain either finite impulse response (FIR) or infinite impulse response (IIR) filters.

The wavelets forming a continuous wavelet transform (CWT) are subject to the uncertainty principle of Fourier analysis respective sampling theory: Given a signal with some event in it, one cannot assign simultaneously an exact time and frequency response scale to that event. The product of the uncertainties of time and frequency response scale has a lower bound. Thus, in the scaleogram of a continuous wavelet transform of this signal, such an event marks an entire region in the time-scale plane, instead of just one point. Also, discrete wavelet bases may be considered in the context of other forms of the uncertainty principle.

Wavelet transforms are broadly divided into three classes: continuous, discrete and multiresolution-based.

Comparisons with Fourier Transform (Continuous-time)

The wavelet transform is often compared with the Fourier transform, in which signals are represented as a sum of sinusoids. The main difference is that wavelets are localized in both time and frequency whereas the standard Fourier transform is only localized in frequency.

The Short-time Fourier transform (STFT) is more similar to the wavelet transform, in that it is also time and frequency localized, but there are issues with the frequency/time resolution trade-off.

Wavelets often give a better signal representation using Multiresolution analysis, with balanced resolution at any time and frequency.

The discrete wavelet transform is also less computationally complex, taking O(N) time as compared to O(N log N) for the fast Fourier transform. This computational advantage is not inherent to the transform, but reflects the choice of a logarithmic division of frequency, in contrast to the equally spaced frequency divisions of the FFT(Fast Fourier Transform).

It is also important to note that this complexity only applies when the filter size has no relation to the signal size. A wavelet without compact support such as the Shannon wavelet would require O(N^2). (For instance, a logarithmic Fourier Transform also exists with O(N) complexity, but the original signal must be sampled logarithmically in time, which is only useful for certain types of signals.)

Definition of a Wavelet

There are a number of ways of defining a wavelet (or a wavelet family).

Scaling Filter

An orthogonal wavelet is entirely defined by the scaling filter- a low-pass finite impulse response (FIR) filter of length $2N$ and sum 1. In biorthogonal wavelets, separate decomposition and reconstruction filters are defined.

For analysis with orthogonal wavelets the high pass filter is calculated as the quadrature mirror filter of the low pass, and reconstruction filters are the time reverse of the decomposition filters.

Daubechies and Symlet wavelets can be defined by the scaling filter.

Scaling Function

Wavelets are defined by the wavelet function $\psi(t)$ (i.e. the mother wavelet) and scaling function $\phi(t)$ (also called father wavelet) in the time domain.

The wavelet function is in effect a band-pass filter and scaling it for each level halves its bandwidth. This creates the problem that in order to cover the entire spectrum, an infinite number of levels would be required. The scaling function filters the lowest level of the transform and ensures all the spectrum is covered.

For a wavelet with compact support, $\phi(t)$ can be considered finite in length and is equivalent to the scaling filter g.

Wavelet Function

The wavelet only has a time domain representation as the wavelet function $\psi(t)$.

For instance, Mexican hat wavelets can be defined by a wavelet function.

Applications of Discrete Wavelet Transform

Generally, an approximation to DWT is used for data compression if signal is already sampled, and the CWT for signal analysis. Thus, DWT approximation is commonly used in engineering and computer science, and the CWT in scientific research.

Wavelet transforms are now being adopted for a vast number of applications, often replacing the conventional Fourier Transform. Many areas of physics have seen this paradigm shift, including molecular dynamics, ab initio calculations, astrophysics, density-matrix localisation, seismology, optics, turbulence and quantum mechanics. This change has also occurred in image processing, blood-pressure, heart-rate and ECG analyses, DNA analysis, protein analysis, climatology, general signal processing, speech recognition, computer graphics and multifractal analysis. In computer vision and image processing, the notion of scale-space representation and Gaussian derivative operators is regarded as a canonical multi-scale representation. One use of wavelet approximation is in data compression. Like some other transforms, wavelet transforms can be used to transform data, then encode the transformed data, resulting in effective compression. For example, JPEG 2000 is an image compression standard that uses biorthogonal wavelets. This means that although the frame is overcomplete, it is a *tight frame*, and the same frame functions (except for conjugation in the case of complex wavelets) are used for both analysis and synthesis, i.e., in both the forward and inverse transform.

A related use is that of smoothing/denoising data based on wavelet coefficient thresholding, also called wavelet shrinkage. By adaptively thresholding the wavelet coefficients that correspond to undesired frequency components smoothing and/or denoising operations can be performed. Wavelet transforms are also starting to be used for communication applications. Wavelet OFDM is the basic modulation scheme used in HD-PLC (a powerline communications technology developed by Panasonic), and in one of the optional modes included in the IEEE P1901 draft standard. The advantage of Wavelet OFDM over traditional FFT OFDM systems is that Wavelet can achieve deeper notches and that it does not require a Guard Interval (which usually represents significant overhead in FFT OFDM systems).

History

The development of wavelets can be linked to several separate trains of thought, starting with Haar's work in the early 20th century. Notable contributions to wavelet theory can be attributed to Zweig's discovery of the continuous wavelet transform in 1975 (originally called the cochlear transform and discovered while studying the reaction of the ear to sound), Pierre Goupillaud, Grossmann and Morlet's formulation of what is now known as the CWT (1982), Jan-Olov Strömberg's early work on discrete wavelets (1983), Daubechies' orthogonal wavelets with compact support (1988), Mallat's multiresolution framework (1989), Nathalie Delprat's time-frequency interpretation of the CWT (1991), Newland's Harmonic wavelet transform (1993) and many others since.

Timeline

- First wavelet (Haar wavelet) by Alfred Haar (1909)
- Since the 1970s: George Zweig, Jean Morlet, Alex Grossmann
- Since the 1980s: Yves Meyer, Stephane Mallat, Ingrid Daubechies, Ronald Coifman, Victor Wickerhauser.

Wavelet Transforms

A wavelet is a mathematical function used to divide a given function or continuous-time signal into different scale components.

Usually one can assign a frequency range to each scale component. Each scale component can then be studied with a resolution that matches its scale. A wavelet transform is the representation of a function by wavelets. The wavelets are scaled and translated copies (known as "daughter wavelets") of a finite-length or fast-decaying oscillating waveform (known as the "mother wavelet"). Wavelet transforms have advantages over traditional Fourier transforms for representing functions that have discontinuities and sharp peaks, and for accurately deconstructing and reconstructing finite, non-periodic and/or non-stationary signals.

Wavelet transforms are classified into discrete wavelet transforms (DWTs) and continuous wavelet transforms (CWTs). Note that both DWT and CWT are continuous-time (analog) transforms. They can be used to represent continuous-time (analog) signals. CWTs operate over every possible scale and translation whereas DWTs use a specific subset of scale and translation values or representation grid.

There are a large number of wavelet transforms each suitable for different applications:

- Continuous wavelet transform (CWT)
- Discrete wavelet transform (DWT)
- Fast wavelet transform (FWT)
- Lifting scheme & Generalized Lifting Scheme
- Wavelet packet decomposition (WPD)
- Stationary wavelet transform (SWT)
- Fractional wavelet transform (?)

Generalized Transforms

There are a number of generalized transforms of which the wavelet transform is a special case. For example, Joseph Segman introduced scale into the Heisenberg group, giving rise to a continuous transform space that is a function of time, scale, and frequency. The CWT is a two-dimensional slice through the resulting 3d time-scale-frequency volume.

Another example of a generalized transform is the chirplet transform in which the CWT is also a two dimensional slice through the chirplet transform.

An important application area for generalized transforms involves systems in which high frequency resolution is crucial. For example, darkfield electron optical transforms intermediate between direct and reciprocal space have been widely used in the harmonic analysis of atom clustering, i.e. in the study of crystals and crystal defects. Now that transmission electron microscopes are capable of providing digital images with picometer-scale information on atomic periodicity in nanostructure of all sorts, the range of pattern recognition and strain/ metrology applications for intermediate transforms with high frequency resolution (like brushlets and ridgelets) is growing rapidly.

Fractional wavelet transforms are based on scaling functions which, contrary to the "integer" wavelet transform, may have infinite support. Nonetheless, they allow non truncation of basis functions, exact treatment of boundaries, and perfect reconstruction.

Imagery Analysis

Imagery analysis is the extraction of useful information from bi-dimensional graphic formats, including screen shots. This includes colour and black-and-white photographs, infra-red photographs and video, radar screens and synthetic aperture radar formats, ultrasound, EKG, EEG, MRI, echo cardiograms, seismographs and others. In short, any type of sensor-related data projected in 2- and 3-D formats qualifies as imagery.

Origins

Prior to the invention of early photography, military commanders depended on scouts that would survey or recon enemy activity, depending on simple eyesight and human memory. Once photography became available, tactical information became frozen in time, details could be preserved, enhancing the quality of available information.

World War I saw the start of ground-based and aerial photographic collection. For the first time commanders were able to access timely and accurate intelligence. Such was the value of this type of information that observers in tethered balloons and scout planes were attacked, first with crude weapons and later escalating to machine guns and the development of the fighter aircraft.

Frank Luke, an American pilot procured incendiary ammunition and used it to destroy numerous enemy observation balloons, gaining the title of *Balloon Buster*. The end of the war resulted in the scaling down of tactical and strategic capabilities, resulting in an almost dormant state in the development of photographic analysis. The perceived threat from Germany and Japan revived the collection and analytical capabilities of the major powers and helped military planners including U.S. General Dwight D. Eisenhower prepare for the next war.

In the 1930s experiments with film media and its processing resulted in the introduction of film that could now detect non-visible wavelengths in the infra-red spectrum. One of the first applications was the use by those associated with rare art collections. Previously invisible details made it possible to detect and deter forgeries.

Radar made its appearance during World War II, primarily in its early warning capability. In the early days of the Cold War, Soviet troops would use a directional radar beacon to lure surveillance aircraft toward their airspace in order to shoot them down. By this time radar scopes became available in larger aircraft monitoring Soviet-controlled border areas. Having these scopes made early radar navigation possible, indeed, in photos released by the Soviet air force, pictures were taken of the screens, documenting this use.

The importance of tactical information is shown in the case of Operation Market-Garden, the aerial invasion of the Netherlands on Sep 17, 1944. Photo missions revealed the presence of two Panzer Divisions in the city of Arnhem, a bridgehead at the farthest reach of those airborne troops assigned.

British intelligence Major Brian Urquhart warned his commander of the threat, but an overpowering optimism cause by the recent collapse of the Western front overruled any possibility of an objective threat assessment, resulting in a night-time river crossing in which out of 10,000 members of the British 1st Airborne Division that jumped into Arnhem, only 2,600 survivors would reach the southern shore 9 days later. In spite of the introduction of colour film, photo interpreters to this day continue to use black and white because of the greater detail available. The early Cold War era also saw the introduction of strategic collection. In tactical collection, analysts count guns; strategic collection includes butter. The categories of collection is, of course, classified.

The post-Vietnam era saw the introduction of airborne infra-red sensors. Differences in temperatures between objects and their surroundings made it possible to detect targets on the ground. These early systems would record data which would be accessed once the aerial platform would land. Later developments in transmission technology would provide periodic data dumps and would further evolve into real-time collection. Synthetic aperture radar would soon be developed in the later part of the Cold War. The concept of an optical camera aperture affecting the image acquisition process would be emulated with radar waves, providing an undisclosed amount of detail. One clue would be the NASA photo released in the late 1980s showing a previously hidden African dry riverbed.

This is also about the time when ultrasound would make its appearance. For the first time it was possible to view variations in tissue density which made it possible to detect possible tissue and organ anomalies. Another application was that of detecting material flaws in manufacturing.

Computer Analysis

Experimentation with monochromatic imagery (black and white) revealed the potential of exploiting the hundreds of shades of grey available in this medium. The next step would be to digitally manipulate the greyscale to enhance the acquisition of usable information. The first applications of this new technology would be by the intelligence community and by medical researchers who would refine and further develop the technology, resulting in the introduction of the CAT Scan.

Another technology introduced roughly at this time would be that of the echo-cardiogram, which could show heart movements and the actual blood-flow through its chambers.

One of the more recent developments has been that of magnetic resonance imaging (MRI), where tissue and blood-flow anomalies could be detected. Evidence of spinal cord injuries and even complex neurochemical reactions in the brain could now be detected and documented. Scientists have also explored the possibilities of multi-spectral imaging such as the 1970s LandSat, and yet more parts of the electromagnetic spectrum such as astronomical gamma-ray imaging.

Analytical Techniques

The first use of tactical imagery obtained during the first World War readily revealed the straight man-made lines of roads, cites, airfields and trenches. Finding concealed high-value targets like artillery, ammo dumps, and other logistical sites was quite another matter. This was a process that was strictly by trial and error, with the resulting body of knowledge transmitted to new recruits and officers. Terrain and the proximity to supported units would dictate probable locations of logistical routes, ammo dumps, supply depots and assembly areas. Being that the military by definition embraces uniformity, patterns of emplacement and concealment, once discovered would result in widespread targeting by artillery and air strikes. The size, shape, and surroundings of items frequently gave away the location of military assets, with shadows only making it that much easier to identify targets. The development of analytical techniques is really a part of the evaluation of the new technology itself. The first photograph to be taken was that of a French neighbourhood. It was crude, yet it clearly showed the outline of the houses. Immediately it was apparent how the new technology, the chemical film plate, was of immediate usefulness.

In the case of infra-red photography, the new details made available were puzzling at first, and took some time to explain. In the pictures taken of works of art, the strange images would eventually be interpreted as showing a feature being painted over and finished. Simultaneous aerial coverage by photo and IR of a given target would reveal how a warm vehicle would warm up the ground and once moved, the warmed plot would stay warm for some time, giving the illusion of more vehicles. Just as in the case of an experienced scientist, once a new observation is made, it must then be explained.

In the case of the application of radar, all there was at the beginning was a variation of the cathode ray tube which would show only the distance to a single target. Only with the introduction of the more familiar round-screen format would radar reach its full potential. So, there were the raw data, but without the use of a readable 2- or 3-D format no-one can make that much use of this information. One thing to remember about radar is that when it comes to illuminating aircraft, most of the energy is deflected. Only the existence of corners, air intakes and flat surfaces that face the radar makes it possible to detect these aircraft. What is actually seen by traffic controllers is the return beep from the aircraft's IFF. As in the case of 9/11, once the hijacked aircraft's IFF was turned off, there wasn't much to see. This can also be seen in the use of radar reflectors that are routinely added to power lines in order to avoid crashes by low-flying aircraft. The actual characteristics of synthetic aperture radar is of course, classified, so one can only speculate on what is actually observable.

For the development of CAT scans, computer-aided design (CAD) had to come first. Pictures were publicized in the 1960s showing design engineers using light pen peripherals to draw proposed design features to be evaluated for fit and aerodynamics before costly manufacturing jigs had to be built. In the case of CAT scans, the information from x-rays is useless without 3-D capability.

For the development of ultrasound, the use of anatomical studies, dissections, and autopsies would have been necessary to provide insight and confirmation of what was now visible. It would have taken some time to establish average dimensions for organs and, in the case of pre-natal scans, body dimensions and growth rates.

The development of MRI would have been a question of comparing their data with that of CAT scans and ultrasound. As far as how they established the visibility of neurochemical reactions, that would have been dependent on current knowledge of neurological and physiological processes. Now a situation exists where a new technology that is based on previous understanding actually increases those fields of knowledge that made it possible.

The current emphasis of multi-spectral imaging is really a question of maximizing the amount of data available for geological, agricultural, and environmental research. This means that a given area would only have to be covered once, making global coverage a more economical proposition.

The latest imaging technologies are driven by nuclear physics and astronomic research. This can be seen in the evaluation of particle acceleration, where theoretical physics helps to make sense of the collected data. As in the case of particle physics, multi-spectral orbital imaging is driven by theoretical research, only to be confirmed by other sources.

Current Applications

Besides the traditional tactical and strategic use by civilian and military intelligence, other entities have made extensive use of this discipline. Law enforcement has made use of imagery in forensic crime scene documentation in order to determine how crimes were committed to include how the assailant approached and left the crime scene. Also, bullet trajectories can be detected in order to determine the location of a sharpshooter.

The United States Border Patrol have the use of imaging technology, determining transit routes and the detection of illegal aliens trying to escape into the interior, beyond the reach of the agents. Their only real problem is that there are far too many routes to cover with the manning and technology only able to do so much.

Highway departments make use of stereo and terrain analysis techniques to determine potential highway routes. As in the case of currently available programs, imagery is included with other types of information to create detailed maps useful for commerce, taxation, city planning, and infrastructure.

The most important application has been for medical and research purposes. Many advances in diagnostics and monitoring have contributed to the ever-increasing body of knowledge and treatment options. The only problem is that with the increase in diagnostic capability, the aspect of accountability and malpractice has made necessary the costly regimen of multiple-discipline testing. This is not about to change. The positive side of developing new imaging technologies is that enhanced observation and understanding will result in better diagnostics and treatments.

The introduction of LandSat in the mid '70s made possible new applications in the fields of agriculture, geology, mining, and the environment. The actual resolution would not be great, but sufficient for these types of applications. The raw data would include the grey scale, and information from a variety of sensors. The designers would find it necessary to assign colours for each type of return, creating a multicolored map.

Meteorological imagery since the '60s has made it possible to detect and monitor severe weather well in advance of its arrival, saving numerous lives.

Future Applications

One promising application would be in the field of archaeology. Terrain analysis would show trade routes, lines of communication, cities, forts, farming, grazing, water sources, supporting communities that surround cities and service trade routes, ancient borders, and more.

In the case of Ancient Egypt, IR would reveal water sources that would have supported communities in the desert. Terrain analysis reveals that in order to access the Sinai copper mines, one had to access the shallow eastward valley north of present-day Cairo and reach the Red Sea just south of Port Said. From there it would have been a question of sailing east toward the western coast of the Sinai and turn southward toward Ras Abu Rudeis, a small coastal plain just east of the two copper mines. The reason for this is that an overland route would have required the costly logistical support of garrisons through territory held by hostile desert tribes.

In the case of the biblical Exodus, terrain analysis excludes the traditional sites as being too far and not being accessible to such a large group of people. Advancing through mountainous terrain would have exposed them to ambushes. The only confirmed location within Egypt or the Sinai is that of Baal Zephon. Ancient papyri describe this location as being close to Ramses, Tahpanhes and present-day Lake Menzaleh. Being that Biblical Archaeology is almost devoid of independent confirmation, one has to use what little confirmed information is available. Following terrain, they would have set out eastward along the Mediterranean coast, reaching the Wadi of Egypt (Al-Arish), and turning southward, following the wadi towards the interior. There are numerous dams crossing the wadi, easily seen from above. Travel would have depended on the use of scouts who would survey water sources, grazing areas and topography that would permit travel for such a large group of people.

Imagery would also benefit exploration in greater Palestine. Radar would readily detect tells (mounds indicative of multiple layers of ruins) in the plains. In mountainous terrain, it would be a question of branching out from confirmed locations and establishing a 10-mile radius, the idea being that cities depend on smaller, surrounding communities. Terrain would dictate probable trade routes, water sources, grazing, farming, and supporting infrastructure. Surveying jungles would require terrain analysis and radar to detect stone cities and temple complexes.

Seismoelectrical Method

The seismoelectrical method (also called the electroseismic method or seismo-electric]) is based on the generation of electromagnetic fields in soils and rocks by seismic waves. Although the method is not reported to detect groundwater flow, it does measure the hydraulic conductivity, which is related to permeability and, therefore, to the potential for groundwater flow.

Operation

When a seismic wave encounters an interface, it creates a separation at the interface forming an electrical dipole. This dipole radiates an electromagnetic wave that can be detected by antennae on the ground surface.

As the seismic (P or compression) waves stress earth materials, four geophysical phenomenon occur:

1. The resistivity of the earth materials is modulated by the seismic wave;
2. Electrokinetic effects analogous to streaming potentials are created by the seismic wave;
3. Piezoelectric effects are created by the seismic wave; and
4. High-frequency, audio- and high-frequency radio frequency impulsive responses are generated in sulfide minerals (sometimes referred to as RPE).

The dominant application of the electroseismic method is to measure the electrokinetic effect or streaming potential (item 2, above). Electrokinetic effects are initiated by sound waves (typically P-waves) passing through a porous rock inducing relative motion of the rock matrix and fluid. Motion of the ionic fluid through the capillaries in the rock occurs with cations (or less commonly, anions) preferentially adhering to the capillary walls, so that applied pressure and resulting fluid flow relative to the rock matrix produces an electric dipole. In a non-homogeneous formation, the seismic wave generates an oscillating flow of fluid and a corresponding oscillating electrical and EM field. The resulting EM wave can be detected by electrode pairs placed on the ground surface.

Surface seismic sources and measurement electrode pairs are generally used to measure the electrokinetic effect. Borehole systems also have been recently developed. A hammer blow or small explosive (black powder) are typically used for the seismic source. Two, short (few meters) electrode pairs are typically located collinear and symmetric to the seismic source. Stacking of repeat seismic "shots" is often used to improve signal to noise.

The only company found that produces a commercially available electroseismic prospecting system is GroundFlow Ltd. in the UK. This system has been mainly used for field demonstration purposes and research and development studies. Very few detailed case histories are available to web manual the performance of the system. The electroseismic system made by GroundFlow is called "GroundFlow 3000" and the main equipment components are:

1. A combined computer-receiver
2. Antenna cables and electrodes,
3. Trigger cable.

The computer-receiver contains a preamplifier, analog to digital converter and power supply system to measure the voltages induced in the grounded dipoles.

The hammer and plate seismic source is used along with two pairs of electrodes arranged in a straight line with electrodes offset (from the centre of the array) by 0.25 and 2.25 m. The seismic source is positioned in the centre of the array. Larger electrode spacing can be used; however, they are typically centred in pairs around the shot point and are generally less than about 10 m in length. Timing of the measurement is achieved by a hammer trigger (or other mechanical trigger). Measurements are made for a period of 400 ms after triggering. The last 200 ms of the record is used as a sample of background noise and is subtracted from the first 200 ms of signal in order to remove the first, third, and fifth harmonics of the background noise at the receivers. This noise is typically from power lines.

The Seismoelectric Method- examples of Successful Field Studies

The propagation of seismic waves in porous rocks is associated with a small transient deformation of rock matrix and pore space which can cause electromagnetic fields of observable amplitude if the pores are saturated. Seismoelectric field measurements are expected to help localize permeable layers in porous rocks and provide information about anelastic properties such as hydraulic permeability or fluid chemistry. This theoretical potential for hydrogeological applications, however, is so far confirmed only by a very limited number of reliable field studies. In addition, these studies usually show only one example of a successful detection of seismoelectric signals. As a consequence, the seismoelectric method is still far from being routinely used. We present a number of field data examples which prove a successful detection of the seismoelectric wave converted from seismic compressional waves at a layer boundary. Our data include the mapping of a 2D-structure. Seismic, radar or borehole data confirm the validity of the seismoelectric results at several measurement areas, which shows that our successful detections are not isolated strokes of luck.

Chapter 7

Fundamentals of 3D Modelling and Visualisation

This chapter presents of an overview of the basic concepts in data modelling, computer graphics and Web techniques, and aims to familiarise the reader with the fundamental concepts and principles used in the thesis and now widely agreed on throughout the GIS and CG society. The chapter can be divided into four sections. First, the chapter clarifies the term *model*, refines its components and discusses the design phases leading to a functional system. Second, the basic principles are discussed in a GIS context. Existing definitions, characteristics and spatial relationships of real objects are commented on. Common object-oriented principles are briefly reviewed in support of the object-oriented concepts used in several chapters of the thesis. The third part is devoted to visualisation techniques and aims to clarify the terminology, basic visualisation mechanism and components to create realistic scenes. The last part discusses common methods to assess, visualise and interact with data on the Web. At the end, the design criteria of VRML and its role in the visualisation process are briefly discussed.

Models and Modelling

The term *model* is one of the most frequently used words in many disciplines. Scientists build and prove hypotheses, make predictions, exchange ideas and gain knowledge on the basis of models. The model is extensively used to represent certain phenomena in a way readable for others. Hence, the meaning of the term, as well as the principles on which to build models, have to be understandable for everyone.

Many different models exist about various aspects of the world. One categorisation of models can be made regarding the phenomenon to be described, i.e. *real* or *imaginary*. In the first case, some properties of the real object have to be selected (measured, investigated, observed) and utilised to create the model. In the second case, the properties have to be designed (planned, computed, decided) by the user and assembled to form the model. A clear boundary between the two models is sometimes difficult to establish. The deficiency of real measurements might be so high that human design is necessary to represent the phenomenon. In both cases, the process of model production is called *modelling*.

Another categorisation of models addresses their form, i.e. *digital* and *non-digital* (e.g. paper maps). Although sometimes better understood by humans, non-digital models cannot be managed by computers. The benefit of using computers for modelling needs no explanation. Models that are intended to interpret the world in a certain way understandable to the computers are called *data models* (Vossen 1991). In general, data models are simpler than most of the scientific models.

Objects, Attributes, Relationships, Operations

A strict definition of a data model can be found in any book about databases. Here, we adopt the one presented in Tsichritzis and Lochovsky 1982. A data model denoted by M is a tool that provides an interpretation of the world and consist of *generating rules* denoted by G and *operations* denoted by O, i.e. $M(G,O)$. The generating rules reflect two aspects of the model: 1) *structure specifications* (G_s) and 2) *constraint specifications* (G_c), i.e. $G(G_s, G_c)$. Structure specifications establish the type and organisation of data, while constraint specifications impose any compulsory model limitations. For example, a real building might be represented in a model as a set of planar, rectangular faces (due to the structure specifications); however, the faces must intersect only along their boundaries (due to constraint specifications). The generating rules operate on data, which are commonly understood as a tuple of objects (categories), their properties and mutual relationships captured in a certain time moment.

The operations describe all the actions that can be performed on the data as some data models allow standard languages for data manipulation to be established (e.g. SQL). Traditionally, five *generic* operations are distinguished: *set currency* (establish a logical position in the model representation), *retrieve* (make available some data to the user), *insert*, *delete* (remove data from the database) and *update* (change existing data). Common *supporting* operations are *selection*, *navigation* and *specialisation*. *Selection* is an operation that allows a number of data to be identified on the basis of three properties, i.e. *logical position* (first, last, prior), *value* of the content or *relationship*. For example, the user may want to see the first 10 (with respect to the identification number) buildings. *Navigation* is an operation that permits a logical path on the basis of a selection to be followed. *Specialisation* is a complex operation that allows a new object (category) to be created on the basis of existing ones. For example, the user may need to provoke the creation of a conglomerate called "shopping centre" of several existing buildings. In some spatial models, the operation will be completed by the *Eulers operators* for adding and removing elements from a spatial database, in such a way that the structure remains manifold. The operation specialisation is then realised by the operations of set theory *union*, *intersection* and *deletion*. Indeed, the user of the model can define as many operations as needed for the specific application.

The possibility for the model to provide users with different, independent subsets of data (called *views*) is considered an essential characteristic of certain classes of models. More details about views can be found in Vossen 1991.

Many models have been developed of varying scope in terms of world representations and generating rules and operations. Some models handle very abstract notions (i.e. database models) and aim at accommodating any type of data; others are more application-oriented and operate with specific types of data (e.g. points, lines, faces in spatial data models). The apparent advantage of the common data models is that the structure and operations are independent of the application and hence the management system built around the model in practice serves any application. The application model is described according to the rules of the common model in a *schema*. The term *mapping* is frequently used to describe this process, e.g. "our application model is mapped onto the relational data model". Among the common models only three have become widely used and implemented, i.e. *relational*, *network* and *hierarchical* data models. The three models are products of the logical design and are sometimes referred to as logical models. Since the relational data model is used later in the thesis, the basis principles of only this model are mentioned here.

Design Phases in Modelling

In obtaining a model, three phases can be distinguished: *conceptual*, *logical* and *physical*. Although the phases refer all kinds of models, the terminology is established by research on models of common purpose. Therefore, exhaustive discussions on the design phases can be found in the literature for database management systems.

Conceptual design clarifies which items have to be included in the model: types of objects, their characteristics and relationships. The resulting product of the conceptual design is a conceptual model or *schema*. The organisation of all the information that the user wants to store permanently on the disk, i.e. creating *views* of reality, seems to be a tedious task and requires appropriate abstractions and rules. The research on *conceptual (semantic) models*, which aims at providing tools to describe real phenomena, is quite extensive and has resulted in a number of more or less sophisticated models. One of the most popular, the Entity-Relationship (ER) model operates with *entities*, which have *attributes* and *relationships* among them. An entity can be anything that has some meaning, e.g. a person, a building, a car, a process. Specific characteristics of entities such as names, colour, address, etc., are referred to as attributes. Since the entities are not isolated in the real world–instead they interact or are associated with each other–the notation about relationship is introduced. The relationship can be *one-to-one, one-to-many*, *many-to-one*, *many-to-many* or *IS_A*. The last relationship represents specialisation (to be discussed later in the text). These notations and the graphical representation called *ER-diagram* complete the ER-model. As it can be realised, the model does not support an aggregation mechanism; instead it offers an easy way to derive relational tables for the relational data model.

Logical design specifies the elements of the logical model, according to the schema obtained from the first stage. If the data model is a relational data model, the logical design clarifies the tables, columns per table and keys. The identification of the objects has to be ensured by introducing a unique ID. If ordering is needed, e.g. clockwise orientation of faces, it has to be indicated explicitly since the relational model inherits non-ordering properties from sets. Often the logical model is associated with the term *data structure,* since the structure of the data is already established. Physical design is the last phase, establishing the way of communication with the hardware with respect to the logical model, i.e. set of files stored on the disk. The physical design has to enable the operations for manipulating the model.

These three stages of design complete the design phase of the model, which is still not sufficient for a system to become managable. Yet, user interface for the interaction and manipulation of the model has to be provided to enable utilisation by the user.

Object-oriented Principles

The modelling process requires many assumptions, classifications, abstractions, selections, generalisations, etc., about real objects and their attributes to be performed. Among the variety of different principles, we concentrate on common object-oriented principles, since they are applied in several parts of the thesis.

The object-oriented approach emerged in early 1970 when the first object programming languages (Simula, Smalltalk) were developed. Since that time, the approach has found applications in different areas of human science. The advantage of the approach is the problem-solving strategy encapsulating information, functions and behaviour per object. Thus any problem to be solved is approached from an object (e.g. person, building, street) perspective rather than from a function ("what kind of functions does the system have to perform") or information ("what kind of information has to be supplied") perspective. The approach is built on a number of principles that provide a mechanism to model real objects in more natural way.

The common methods facilitate the organisation of objects' characteristics and their relationships with other objects.

Objects and attributes: An object can be anything, i.e. person, action, process and the *attributes* represent all the characteristics. Methods, processes and functions denoting *behaviour* aim to represent the dynamics of objects. For example, an object can be a building with attributes colour and height. Its behaviour can be the ability to be created, edited, selected, etc.

Wholes and parts: The method has at least three variants: "assembly and parts", "container and contents" and "group and members". The first relation represents the obvious link that exists in manufacturing parts. The door is a *part of* a car, the window is a *part of* a wall, etc. The second variant helps in cases when the "things" are not really designed for each other. A typical example is an office, which *consists of* a desk, a computer, a chair, and bookshelves. The third expression comprises relations that may appear in formations like faculties and courses, groups and subgroups. The members of a subgroup are considered members of the group as well. For example, the group of ITC students *consists of* the students of courses GFM, GIC, GIR, etc. The principle behind the whole-part method is *aggregation*. Each *whole* aggregates a given number of *parts,* i.e. the cardinality (object connection constraint) of relationships has to be indicated. The relationship is usually read from top to bottom, i.e. *consist of.*

Classes and instances: Each object has to be organised in a class as the classes are formed with respect to common attributes and behaviour. Each object of a class is an instance of that class. For example, the ITC building is an instance of class Buildings in Enschede. The mechanism to form classes or specify new objects is known as *generalisation-specialisation*. The generalisation part is applied when a class has to build, i.e. the new class "accepts" (some of) common characteristics of objects and thus creates a more general view about a certain group of objects. The new class becomes their parent and the objects are its children. The children inherit the common properties from the class and contain only the ones specific to them. The specialisation part of the mechanism is used to create children. The number of classes and, respectively, children is not limited. Thus a hierarchy is built as the generalisation mechanism extends the hierarchy upwards and the specialisation mechanism downwards. In through contrast to *wholes-parts,* the *classes-instances* relation is read from bottom to top, i.e. *is a*.

Abstraction is a principle considering the characteristics of an object that are important for the problem or application. Some details quite important for one application might be irrelevant for other. Abstraction is the first principle, which has to be thought for object identification. The characteristics of the objects can be clarified only after an accepted level of abstraction.

Encapsulation is an object-oriented programming mechanism to combine an object characteristic and a corresponding behaviour. Furthermore, the particular software design unit (subroutine, method) could be hidden from the user. For example, a "click" with the mouse on a building (on a monitor) may visualise the owner of the building.

The event "click" detected by the system activates the retrieval of a particular object's characteristic. In this example, the user action "OnClick" and the operation "retrieve owner" are encapsulated in one single method (behaviour). The encapsulation principle applied to the organisation of data can be understood as providing (or not) access to parts of the data and operations. Certain responsibilities can be assigned to parts of the information to restrict operations on data. For example, a common user of a municipality GIS may not be aware of actual geometric representation of a building and consequently not allowed to change the geometry.

Inheritance is the mechanism to express similarities (common characteristics) between objects. As mentioned above, it is the basic mechanism for building classes and thus helps in propagating properties from parent to child. In general, all the attributes and behaviour of a parent are inherited by the children and grandchildren and so forth. The *overriding* and *expending* of parent attributes and behaviours is allowed as well. The case where a child has only one parent is known as *single* inheritance, while multiple parents invoke *multiple* inheritance. These supplementary techniques are useful for creating composite classes. Opponents of *overriding* and *extending,* and *multiple* inheritance say that the simplicity of the model is lost and that operation with overridden attributes is more complicated. An alternative is introduction of *local* and *inherited* properties.

Polymorphism is the ability of objects to have different forms. For example, an object changing its geometry from very detailed to a simple cube is a morphing object. A bank transfer may settle a tax payment by filling out a form or by e-mail. All the operations are valid, therefore the tax payment is considered polymorphic. Polymorphism is a widely used mechanism to speed up real-time navigation inside 3D models. Each object or groups of objects have several geometric representations and the visualisation software decides which of them has to be used. Attributes of objects can be polymorphic as well. For example, an attribute document number can be associated with a parcel as a property document and a person as a document for ownership.

Association is a mechanism to establish a connection between the objects in two classes. The relationship is not typical for the entire class and in this sense can not be resolved by aggregation. For example, class Building and Street can have a connection concerning the relation "building on a street". This leads to the construction of a new class Building-on-Street. Either of the two classes Building and Street is a *part of* the new class, therefore no aggregation can be applied. However, the cardinality must be known. The similarity between the aggregation and association mechanisms is quite high and sometimes it is hard to decide which mechanism to apply. The basic rule is that aggregation is a class connection pattern, while association is object connection one. Several associations have become well known and used, and have their own name, i.e. *participant-transaction*, *place-transaction*, *participant-place*, *transaction-transaction*, *peer-peer* (object connection between objects in he same class).

Relational Data Model

The relational data model based on *relations* and their representation as tables was introduced first by Codd. The basic category relation is defined as follows: R is a *relation* on given sets $D_1, D_2,...,D_n$ if it is a set of *n-tuples*, each of which has its first element from D_1, second element from D_2 and so on. The sets $D_1, D_2,...,D_n$ are called the *domains* of R. The number of the domains is the *degree* of the relation and the number of the tuples in R is its *cardinality*. A table represents each relation, and each column of a table is called *attribute*. One or more relations, then, can define an object type. Thus the database is specified by a *relational schema*, which consists of one or more relations R. Since the relational schema consists of separate tables, the relationships between the tables are established by a propagation of *keys*. The key can be composed by one or more attributes. Two constraints must be satisfied: the key must be unique and the key must be minimal. For example, the identification number (ID) of a building can be used as a key. Since representation by tables may easily lead to repetition of information, the optimisation of the model is critical. *Normalisation* is a procedure to eliminate different types of dependencies among the relations and attributes. Theoretically, five normal forms are defined, however, three of them are commonly considered sufficient. The basic concept that the relational data model misses is the existence of constraints on and among the relations. For example, it is impossible to specify that certain types of buildings (e.g. residential) cannot be higher than four floors.

Among the many different languages for data definition and the manipulation of relational data models, the Structured Query Language (SQL) enjoys the greatest popularity. SQL is an English-keyword set-oriented language. The basic concept is that each operation is *mapped* onto the relation. The effect is similar to scanning the columns of a table and checking for a value or a set of values. For example, for the operation retrieval, all the rows of the table are scanned and the ones where the sought value is found are returned as results. The fundamental set of key words SELECT–FROM–WHERE allows a large number of mappings to be performed. Furthermore, several mappings can be combined to ensure the basic set operations *union, intersection* and *difference*. Attempts to extend SQL toward an *ad hoc* language for a GIS have been made without a significant success. Egenhofer 1992 argues that SQL cannot become a GIS language due to a number of impossible operations: identity queries, metadata queries, knowledge queries (explaining the reasoning), qualitative queries (e.g. which road is wider), display of query, modifying the database content from the graphics and modifying the graphical presentation. The inadequacy of SQL is currently overcome by utilising *embedded* SQL queries. The outcomes of SQL queries are further processed by operators written in standard programming languages (C, Pascal, Perl) called *host* languages.

Despite all the deficiencies of the relational data model and thanks to the simple concepts and the standard query language, many commercial Database Management Systems (DBMS) have adopted the model that contributed to its successful integration in commercial and administrative applications.

GIS Models

Geo-science specialists are busy with the modelling of real phenomena. A universal model to comprise all the aspects of reality is not practically realisable due to the high complexity of the real world. Different disciplines emphasise different aspects and only these aspects are included in the model. Thus a model considered good for the description of particular phenomena might be hardly appropriate for anothers. Different aspects and characteristics of real objects have led to the existence of several variations in object definition.

The term *GIS model* is utilised here to denote a data model of the real world. Being a data model for representing real-world objects, the GIS models have the components defined above, i.e. object types, relationships and attributes with corresponding generation rules, constraints and operations defined with them. A GIS model with the corresponding user interface constitutes a 3D GIS. Often the GIS content is specified as data about geometric (shape, size, location) and semantic characteristics (called attributes), spatial relationships and time. The *functionality* of GIS is then said to be the possibility to perform operations on data in order to analyse them.

The following sections present the necessary fundamental concepts related to the representation of objects, attributes and relationships in a GIS context.

Types of Objects

Since the interest in geo-sciences has traditionally been directed toward real objects with spatial extent, the differentiation between *spatial* and *non-spatial* objects is widely accepted. A spatial object stands for a real object having *geometric* and *thematic (semantic)* characteristics. Tempfli 1998c draws attention to a third group, i.e. *radiometric* characteristics of spatial objects. Current GIS models maintain only spatial objects. Non-spatial objects (organisations, departments, people, goods, etc.) either are the concern of Database Management Systems (DBMS) or are integrated in GIS as semantic characteristics of spatial objects. For example, the person, who owns a building (spatial) exists in a GIS as a semantic characteristics "owner" of the building. As will be discussed in this thesis, dealing with only non-spatial objects may be a drawback for some applications.

A further distinction is made between real objects with respect to distinctness of properties, i.e. objects with well defined spatial extent and properties and objects with unknown or non-well defined spatial extent and properties, i.e. objects with *discernible (determined)* and *indiscernible (undetermined)* boundaries. Usually, quite independent research is conducted in both areas. This may cause problems in the case of analysis where objects from the two groups are needed. Pilouk 1996 discusses the subject in details and advocates an integrated approach while modelling spatial objects.

Many real objects need the monitoring of some of their characteristics with respect to time. For example, growth of population may cause the fast conversion of land from rural to urban use, which can have an impact on the entire urban development. The geo-science specialist might need to store and analyse the changes in order to be able to predict and control the process. Similar problems have created the branch of temporal definitions of an object, according to which objects are subdivided into *spatio-temporal* and *non-spatio-temporal*. Raza et al 1998 present spatio-temporal-attribute objects (STAO) with three fundamental components, i.e. *location (spatial)*, *attribute (aspatial)* and *time (temporal)*. Cheng 1999 defines a real object by its three aspects *geometry*, *theme* and *time*.

More general definitions of real objects, regardless of the nature of the objects can be found in the object-oriented literature. Coad and Yourdon 1991 denote as an object every item of interest (real object, feature, process) and present a framework to investigate and classify the characteristics of the different objects.

Attributes

Attributes stand for the characteristics of real objects, which are important for the GIS model. The term attribute is frequently used to denote only semantic characteristics of objects. This thesis considers a wider meaning for attributes, i.e. they compile semantic, geometric and radiometric characteristics of objects. The geometric characteristics refer to position (and orientation), shape and size of real objects. The radiometric characteristics regard material, colour or reflectance. The semantic characteristics specify status, functionality, "meaning", usage, etc., of the real objects. This thesis concentrates on geometric characteristics and therefore some basic principles for representing them will be mentioned.

The description of geometric characteristics requires a priori clarification of the abstraction of space, the dimension of space and objects, and the method for representation. Conventionally, two approaches to space abstraction are utilised in modelling processes, i.e. field-oriented and object-oriented. The field-oriented approach assumes complete subdivision of the space into smaller, regular partitions, e.g. pixel. In the object-oriented, the space is "empty" and all the objects are places (embedded) in it, i.e. a lake in a 2D map. Both approaches have advantages and disadvantages and are appropriate for different applications. While the field-oriented approach better suits the representation of continuous phenomena, e.g. height fields, rainfalls, the object-oriented approach represents better discrete phenomena, e.g. buildings, roads.

The dimension of space is defined in mathematics as the number n of a sequence of n -real numbers $(a_1, a_2, \ldots, a_n)$, called an ordered n-tuple. All the ordered tuples are called n-space and are denoted by

R^n. The basic idea is using a tuple of points to define a position in the space, e.g. a pair of points (x and y co-ordinates in Euclidean space) define a position in a plane, i.e. 2D space. The real world is usually represented as a 2D or 3D space. Within the space, the objects have their own dimensionality, i.e. they can be represented by one of the following generic types: 0D (points), 1D (lines), 2D (surfaces) and 3D (bodies) objects. The decision as to the representation of real objects (i.e. point, line, surface or body), is highly influenced by the purpose of the model.

The possible geometric representations are split into three large groups: i.e. raster, vector and constructive primitives. The "building blocks"(named constructive objects in this thesis) in the raster method are regular cells, e.g. pixels (in 2D space), voxels (in 3D space), which fill in the entire object. The representation is simple and easy to maintain, but produces a lot of data for storage, and the overall precision is low. The vector method is based on irregular n-cells, where n={1,2,3} composed of points with co-ordinates. In contrast to the raster method, the cells represent the boundaries of the objects. The last representation, known as Constructive Solid Geometry (CSG), uses irregular 3-cells. An object is represented by a CSG-tree, which holds information about the CSG primitives and the operations "gluing" them (i.e. Boolean operations). The approach is widely implemented in the manufacturing industry. In general, the irregular cells describing the boundaries of the objects allow more precise descriptions of shapes and spatial relationships compared to the raster cells and CSG primitives. Moreover, the visualisation algorithms operate with the boundaries of objects, i.e. CSG and raster representations have to be further processed to determine the boundary. Therefore many applications give preference to the vector approach of description. This thesis is based on a vector representation too.

The vector representations are often referred to as boundary representations (B-rep) or surface representations. A large number of spatial models are developed and implemented in GIS, CG and CAD systems based on irregular multidimensional cells. The names and construction rules of the cells in the different models usually vary. The simplest set of cells is the set of simplexes, i.e. 0-simplex (node), 1-simplex (arc), 2-simplex (triangle) and 3-simplex (tetrahedron). Composing simplexes, one can obtain more complex objects, i.e. simplicial complexes. Most of the models allow 1,2,3-cells with an arbitrary shape that imposes some supplementary constraints, e.g. planarity of faces, convex shape. The names vertex (point), edge, face (polygon) and solid (polyhedron) are then used in the literature to denote 0,1,2,3-cell.

Spatial Relationships

The particularity of GIS compared with other information systems is the maintenance of spatial relationships, i.e. the connections or interrelations between real objects in the geometric domain. The aspects of the spatial relationships are currently under investigations, i.e. approaches to representation, naming and equivalence.

Three different approaches to encoding spatial relationships are discussed in the literature, i.e. *metric*, *topology* and *order*. The metric is a pure computational approach, based on the comparison of numerical values related to the location of the objects in the space (i.e. the Euclidean space). For example, the spatial relationship between a house and a parcel (e.g. inside, outside, to the south) can be clarified by a metric operation point-in-polygon performed for each point constituting the footprint of the building. Since the metric is built on the notion of the *distance function*, which is dependent on the internal (finite) representation of numbers in the computers, the approach is computational expensive.

The order establishes a preference based on the mathematical relation "<" (*strict order*) or "£ " (*partial order*), which allows an organisation of objects similar to a tree. For example, if a building is inside a parcel, the spatial relationship is represented as "building < parcel". The applicability to representing spatial relationships is investigated by Kainz 1989 who argues that it has advantages in expressions of *inside/outside* relationships.

Topology allows the encoding of spatial relationships based on the *neighbourhoods* of objects regardless of the distance between them. The main property of topology, i.e. the invariance under topological transformations (i.e. rotation, scaling and translation) makes it appropriate for computer maintenance of spatial relationships. The following section discusses some basic topological issues that the thesis utilises.

Detection of Spatial Relationships

The development of a mathematical theory to categorise relations among spatial objects has been identified as an essential task to overcome the diversity and incompleteness of spatial relationships realised in different information systems. The intensive research in this area has led to the development of a framework based upon set theory and general topology principles and notions. The framework utilises the fundamental notions of general topology for topological primitives to investigate the interactions of the spatial objects. The topological primitives of a spatial object can be defined for each spatial model and hence the framework can be applied to any spatial model. The basic criterion to distinguish between different relations is the detection of empty and non-empty intersections between topological primitives. Depending on the number of the topological primitives considered, two *intersection models* were presented in the literature. The first idea is to investigate the intersection of interiors and boundaries of two objects. This results in 2^4=16 relations between two objects. Eight relations are given names, i.e. *disjoint*, *meet*, *contains*, *covers*, *inside*, *coveredBy*, *equal* and *overlap*). For example, if the boundaries of the two objects intersect but the interiors do not, then the conclusion is that the objects *meet*. Apparently, many relations cannot be distinguished on the basis of only two topological primitives, therefore the evaluation of the exterior is adopted. The number of detectable relations between two objects increases to 2^9=256. Despite the criticism (i.e. not all the relations are possible in reality, the intersections are not further investigated, many object intersections are topologically equivalent), the framework provides a systematic, easy-to-implement way of detecting spatial relations and is therefore used in the thesis. A slightly different approach is followed by Clementi et al 1993. Again, the three topological primitives are used but first the type of intersection is clarified and then a detailed evaluation of all the cells composing an object is performed. The approach claims a detection of a larger number of relations, however, at a high computational price.

Operations

Apart from the generic operations, GISs perform a number of specialised operations, some of which fall into the group of supporting operations, i.e. selection, navigation and specialisation. Plenty of classifications of required operations are presented in the literature that often confuses the reader. Usually, the classifications try to address three components of GIS, i.e. the geometric and semantic properties, and spatial relationships.

Goodchild 1987 specifies the spatial operations, which have to be performed in six groups:

- operations requiring access to semantic properties of one type of object, e.g. houses that belong to one owner
- operations requiring access to both semantic and location information, e.g. houses higher than three floors located to the north of the station
- operations which create object-pairs from one or more types of objects, e.g. all the buildings included in a given parcel
- operations which analyse semantics of object-pairs, e.g. the buildings in a given parcel that have one owner
- operations which require access to semantic and location information for more than one type of objects or object-pairs: spatial interaction modelling requires access to origin and destination of objects, as well as, their attributes.
- operations which create a new type of object from existing objects, e.g. Thiessen polygons or buffering polygons around lines.

Aronoff, 1995 summarises the required operations of a GIS as:

- retrieval operations, e.g. what is the current information about a particular building
- query operations, i.e. retrieval under a condition e.g. the houses cheaper than $50 000
- modelling operations, e.g. what will be the state of the data after ten years or the next year (predicting new information).

Further elaboration on needed retrieval and query operations results in subdivision into four groups:

- retrieval and query of the spatial data, i.e. transformation (geometric, format, geographical, conflation), and editing (edge matching)
- retrieval and query of the semantic data
- integrated analysis of spatial and semantic data, i.e. retrieval, classification, measurement, overlay operations
- neighbourhood operations (search, topographic operations, Thiessen polygons, interpolation, contour generation)
- connectivity operations (contiguity measurements, buffering, network, spread, seek, inter-visibility, illumination, perspective view)

- output formatting (map annotation, text labels, texture pattern and line style, graphic symbols).

In this thesis, we assume that GIS has to be able to perform selection, network and specialisation operations over the components it contains (e.g. geometric and semantic characteristics, and relationships). The result of these operations can be further processed to obtain a more specific result. Theoretically, the original operation and the further processing can be encapsulated in a new operation. Thus, we will define the following groups of GIS operations that are used in the thesis as follows:

- metric operations, i.e. distance, volume, area and length, are selection operations based on shape and size of objects and further processing
- position operations are selection operations based on location (no further processing)
- proximity operations are selection operations based on geometric characteristics and the creation of new object (i.e. buffer)
- relationship operations are selection operations based on spatial relationships (no further processing)
- visibility operations are a selection based on geometric characteristics and further processing
- semantic operations are selections based on semantic characteristics
- mixed operations, i.e. selections on the basis of geometric and semantic characteristics.

Molenaar 1998 specifies the *query* as a selection operation with three components: data type specification, conditions and operations that have to be performed on the data. The selection then can be performed on semantics, geometry or topology. For example, "select the buildings (data type) higher than 15m (condition) and show their ID (operation)". If the operations on data are not very complex, the definition of a query operation is similar to that given by Aronoff and Goodchild. Sophisticated operations on data may diminish the boundary between query and analysis. Therefore, a selection of data under conditions related to geometric, semantic characteristics or relationships, which are stored in the database, will be referred to here as *ad hoc query* or *query*. A selection of data that needs further processing is called *embedded query*. Embedded queries may be provided to the user as encapsulated new GIS operations.

3D Visualisation and Interaction

Visualisation is a general term to denote the process of extracting data from the model and representing them on the screen. This thesis concentrates on the representation of spatial data and, therefore, visualisation will be mostly used in the context of display of 3D graphics. Interaction with the 3D graphics on the screen is, in the broad sense (details are given below), detecting user actions and reaching them by re-computing the parameters of the models and producing of a new display. Since the 3D visualisation and interaction are critical elements the user interface in a 3D GIS, some fundamental principles (relevant for the thesis) are presented here.

The process of visualisation and interaction is completed in two steps, which are known as *pipelines,* i.e. input and output. The output pipeline comprises the process of sending information (primitives and attributes) to the screen; the input pipeline comprises the detection of user actions and the corresponding post-processing of the model. The visualisation schema is common for any program, with a graphics output, i.e. CAD systems, e.g. AutoCAD, Intergraph, Microstation, GIS programs, e.g. ArcVIEW, MapInfo, Web browsers, e.g. Netscape, Mosaic.

The *application program* is the software controlling and ensuring the correct flow of data in both directions, i.e. 1) it accesses the data organised in a model, extracts a subset of data, which has to be on the screen at one particular moment and applying an appropriate graphics package (rendering engine), performs the necessary algorithms to transform 3D graphics to 2D image on the screen, and 2) it controls the input devices. The process of creating images from the model is often referred to as *rendering*.

The *graphics package* is a software interface to the graphics hardware, which provides means for the rendering, maintaining windows and detecting user actions. The tendency over the last few years to develop hardware-independent software has resulted in a *rendering package,* i.e. OpenGL. From a programming point of view, OpenGL is a set of standard procedures (written in C++) that facilitates the management of graphics hardware. To be able to run on different platforms, the OpenGL syntax does not support procedures dealing with windows and input devices. An application program using OpenGL has to utilise the windowing mechanism and the device tracking (e.g. detecting mouse click) of the current operation system (e.g. Windows, Unix), i.e. make use of corresponding libraries (e.g. Motif X for Unix). In return, OpenGL provides the means: to describe objects, arrange them in three-dimensional space, calculate the colour and illumination parameters and convert the mathematical description of the model to the pixel parameters on the screen. Similar rendering packages, but being less hardware independent, are Direct 3D (Microsoft) and Java 3D (Sun Microsystems), which is the newest 3D interface currently available for two operation systems, i.e. Windows and Solaris.

The visualisation of 3D graphics differs significantly from 2D graphics on several points: 1) scene creation, 2) tools for interaction and manipulation and 3) amount of data. The following text is meant to provide basic knowledge about the mechanisms to obtain readable scenes and ensure sufficient interaction tools.

Scene Components

The word "scene" is used to denote the rendered image on the screen. Several key components are necessary to achieve readable 3D dimensional images on the screen or, in other words, to create realistic scenes: *geometry*, *illumination* and *shading models*, *texture*, *camera position* (or view). The term "geometry" here refers to the representation of geometric characteristics of objects in the scene. The components attempt to capture features of 3D real objects and represent them in accordance with human perception. The *3D geometry*, i.e. *polygons* (*faces*) and *lines*, composed of *vertices* (points with known co-ordinates), determines the position, shape and size of the objects. *Illumination* and *shadingmodels* control lighting from artificial light sources and corresponding reflections on the surfaces. *Colours* and *textures* aim to represent surface properties of the objects. The *camera position* specifies the location of the user inside the model and orients the model accordingly. The readability of the scene depends on the components used to render the model or part of it on the screen. In this respect, three basic techniques can be distinguished, i.e. *points,"empty" polygons* and *"filled"polygons*.

Points and "Empty" Polygons

The scene created only by a set of points is known as *point cloud*. The view gives very few details and only a very vague impression of the entire 3D model that can be obtained. The scene composed of "empty" polygons is known as a *wire frame* mode. A slight improvement in the legibility of data can be achieved by using different colours or by applying algorithms to hide "invisible" lines. Several other more sophisticated techniques, e.g. depth cueing, depth clipping, can be used to improve visual perception. More information about them can be found in Watt 1993. The wire frame mode involves less complicated rendering algorithms (hence much faster) and therefore many rendering engines provide real-time navigation in wire frame mode. However, all the scenes created in wire frame mode, have the common disadvantage of illegible scenes, with a low degree of realism and difficult for orientation. Very often some objects create an ambiguous perception, e.g. it is impossible to determine which elements are closer to the observer.

"Filled" Polygons

The utilisation of "filled" polygons offers an improvement of the realism, which requires the definition of *illumination models* and corresponding *shading models*. The *illumination model* is related to the light sources used and consists of a number of parameters, which determine the colour of every pixel from the rendered scene. A *shading model* is a more broader term, in which the illumination model fits, and defines how the illumination model is applied and what parameters it receives. For example, some shading models implement an illumination model for only a subset of all the pixels of the screen. The values of the remaining pixels are interpolated. Other shading models compute the value of each pixel of the screen. The applied illumination model and shading models can become quite complex if the interchange of lights from the source to the surface and between all the surfaces is considered, e.g. recursive ray tracing and radiosity methods.

The shading model depends on the method of computing the pixel value across the surface. Several approaches are known: 1) constant shading, called *flat* or *facet* – one value is computed and applied to the whole *face*; and 2) interpolated shading: *Gouraund*– computation of intensity in every vertex and applying an interpolation across the surface, *Phong* – interpolation of the normals to surfaces, as the first step is computation of the normals to the edges of the surface.

Complex illumination and shading models to render filled polygons may produce highly realistic scenes. However, they slow down the rendering process tremendously, e.g. applying *radiosity* (a comprehensive shading technique), a scene might take hours to be computed. Some advantages might be found in the little disk space and memory required to process data. In practice, such models are not relevant for real-world models. Shorter time for rendering can be achieved by utilisation of simple *illumination* and *shading models* (i.e. *Gouraund* and *Phong*) or *texturing*. Simple algorithms for shading, however, may create the opposite to a realistic scene. For example, some disturbing effects may occur, like a polygonal silhouette on the edge between two surfaces or non-constant shading of the surfaces (during navigation through the model). It seems that texturing is the most successful technique to obtain realism in a short rendering time.

Texturing is a technique to wrap an image (scanned real photos or synthetic images) onto the "geometry". The principle is such that, instead of the value calculated from illumination and shading models, a corresponding pixel value (or weighted mean value) from the image (texel) is assigned to the screen pixel. Since the screen pixel value pre-computation is not performed, the rendering is faster. The rendering engine, in this case, must cope with the organisation of images in the memory, as well as, the extraction of each new image needed for a particular scene from the database.

Computer graphics standards (e.g. OpenGL, VRML, Java 3D) give preference to simple illumination and shading models combined with methods for texturing. Thus both rapidly rendered and realistic scenes are ensured at the price of additional maintenance of photo images. For example, the virtual reality browsers are expected to provide only *Goraund* and *Phong* models and support texturing.

Interaction and Manipulation

The term *interaction* is related to the possibility to move and investigate (e.g. operations zoom, pan, walk-through, flyover, examine) the model, without changing the original parameters of the scene. *Manipulation* is the ability to access (e.g. the operation select) elements from the scene and the possibility to change (e.g. operations edit, add, delete, scale, rotate, translate) the entire scene or some components of it, i.e. geometry, lights, camera position, image for texturing, texture co-ordinates. The terms have an important role in the visualisation process and are used to classify the visualisation systems and overall visualisation goals. The time needed for a system to respond to a user action is the most critical factor in distinguishing between high or low levels of interaction and manipulation. The time interval gives users the perception of being “inside the model”, which Isdale 1998 uses to distinguish between *off-line plotting*, *interactive plotting*, *animation* and *virtual reality*. *Off-line plotting* is a class of interaction where the user cannot interfere with the process of drawing (e.g. plotting). *Interactive plotting* refers to systems with tools for interaction and manipulation, which, however, need time to re-compute and render the new scene (e.g. most of the CAD systems). *Animation* allows the user to calculate dynamics in advance and play it in real time (e.g. 3DStudio, TrueSpace, Wavefront etc.). Depending on animated properties, four types of animation can be defined, i.e. *frame*, *skeleton*, *morph* and *interpolation*. *Frame* animation gives the ability to plan the route and the position of the camera, and a sequence of scenes (number of frames is defined by the user) is created. *Skeleton* animation changes the positions of some geometric elements of the object, e.g. an arm of a body. *Morph* animation refers to changes in the shape of the object. *Interpolation* means sophisticated computations between two different positions of an object. All the types of animation can be observed but not influenced while they are played.

Virtual reality (VR) is the highest level of interaction and manipulation, permitting operations to be completed in near real time. This means that the user does not notice the intermediate calculations between two different scenes. The manipulation and interaction comprise all facilities (hardware and software) that supply users with tools for contact with the objects and directly manipulate them. The perception of the user of interactivity depends to a great extent on the available hardware. While an equipment based on a conventional *mouse* and 2D screen facilitates only navigation, *Head Mounted Display (HMD)* (a device which allows displays to be mounted in front of every eye) creates true 3D perception. With respect to the hardware available, VR systems can be classified at five levels:

- *Entry VR* systems consist of a personal computer (or workstation) with a 2D input device, 2D screen and video accelerator cards.
- *Basic VR* systems consist of special hardware devices for selection such as PowerGlove, 3D or 6D mouse or joystick, and some enlargement of the display facilities, e.g. stereo graphic viewer.
- *Advanced VR* systems already include all types of rendering accelerators, e.g. texture, video and powerful processors.

- *Immersive VR* systems include complex devices for display and sensing, e.g. HMD, CAVE.
- *Cockpit Simulators (Cabs)* allow physical movements and create a feeling of real participation in the scene and actions.
- Recent hardware and software advances make the realisation of *Entry VR* systems possible on any personal computer. Real-time interactions and manipulations of large models in shaded (or textured) mode are commonly considered low level VR techniques. Systems that provide means for both real-time interaction and manipulation are referred to as *VRmodellers*.

Levels of Detail

The concept of Levels of Detail (LOD) has been introduced to facilitate visualisation of large scenes. The idea is that objects or some of their elements become compatible with the pixel size of the screen when they are far way from the observer. This permits the original geometric representation to be replaced with a new low-resolution one. Low-resolution representations require less memory and processing time for rendering and hence speed-up the visualisation process. The different representation is used by the visualisation system only if the object is far enough from the user. Closer objects are still represented in their full resolution. Moreover, if the distant object gets closer (as a result of the user's navigation through the model), then the high-resolution representation is restored. The intentions are an unnoticeable switch between low and high levels of detail. With some exceptions, the switch of LOD is controlled by one parameter, i.e. the distance between the viewing point and the object.

Most of the algorithms developed assume that the LOD are pre-computed different representations of objects. For example, a building may have four different geometric descriptions: 1) walls, roof facets, windows and doors (i.e. LOD0), 2) walls and roof facets (i.e. LOD1) and 3) the bounding box (LOD2) and 4) a single point, e.g. the mass point (i.e. LOD3).

All four levels have to be computed in advance and available to the system. Research is conducted toward automatic generation of LOD on the basis of previous scenes, which so far gives good results only for scenes similar to each other. Originally proposed for geometry simplification, the idea of LOD might be extended to the images used for texturing. The principle is the same: an image used for a closer object has a higher resolution than the image applied for distant objects. LOD are currently implemented on almost all visualisation systems, but concerning only geometry. VRML specifications present a flexible organisation of geometry and texture LOD based on separate a priori designed VRML documents.

The World Wide Web

Development of the World Wide Web (the Web) can be considered a turning point in data access and exchange over the Internet. Since the first point-and-click browser (Mosaic, NCSA) launched in 1993, the number of Internet customers (individuals, companies, non-profit organisations) has increasingly grown. Several key characteristics contribute to this phenomenon:

- The Web uses a protocol for the exchange of data built for the Internet. The HyperText Transfer protocol (HTTP) is one level higher than the Transmission Control protocol (TCP), which is an extension of the basic Internet protocol (IP), both generally known as the TCP/IP protocol. Since the TCP/IP protocol is equal for all computers, the computers already connected in Internet by Local Area Networks (LAN) and Wide Area Networks (WAN) could adopt the protocol simply by installing a piece of software.
- The Web engine for exploring information is the *browser* or *Web browser,* i.e. a program that works on the client site and supports the most used protocols developed before HTTP, e.g. the File Transfer protocol (FTP), TELNET (text-oriented remote service), Gopher (a university information protocol) and the e-mail message protocol. The user obtains functionality in one single program. The most widely used Web browsers are *Netscape*, *Microsoft Explorer* and *Mosaic.*
- The Web establishes a simple yet effective naming schema, i.e. the Uniform Resource Locator (URL), to identify the unique location of the information and the protocol to obtain it.
- The information on the Web is organised in standard units or *pages.* The page is any document displayed in the browsers and the user can read it. Moreover, applying the *hypertext* mechanism, each page can provide a link to any other document on the Web. Thus every customer of the Web may provide an unlimited number of links to sources all over the Web, which undoubtedly facilitates the exploration of information.

- The Web pages are written in a standard high-level Hypertext Markup Language (HTML). The language gives instruction to the Web browser on how to visualise the document. Although the language specifies many parameters, e.g. font and size of text, the visualisation (e.g. line-breaking) is the responsibility of the Web browser on the user's computer. Thus, the display of the documents is practically platform-independent.
- The Web can be unlimitedly extended with supplementary programs in two directions: on the client and on the server site. On the client site, the Web browser may incorporate different additional programs called *plug-ins*, *helper applications* and *applets,* in order to display specific types of documents. On the server site the Web can be extended by a variety of executable *scripts*.
- The Web browsers provide a user interface based on point-and-click operations that does not require the memorising of commands. Thus, in a very short time, the Web has become popular among societies and individuals with different qualifications.

Techniques to Access Data on the Web

The principle mechanism to access documents on the Web is based on *client-server* architecture. An architecture in which a computer (personal or workstation) called *client*, requests something (information or process) from another supplying computer (personal or workstation), called *server*, both of which are connected in a network (LAN, WAN, Internet), is known as *client-server*. The client has software that is capable of connecting to the server, and requesting and receiving data. The server has software that detects the request for communication, completes the request and closes the connection between the two computers.

Similarly to the standard client-server concept, the Web requires a Web server (e.g. *Apache*, *WebSite*, *WebStar*) on the server site and a Web browser on the client site. The Web browser displays the documents and processes the user actions. The Web server can both establish a connection to other servers or applications by the execution of *scripts* and provide access to local files. The scripts (often called CGI scripts) are programs written according to the standard Common Gateway Interface (CGI) and located in a special directory on the server. CGI scripts, which can be written in any programming language

(Perl, C, Delphi) have three parts, i.e. *header*, *body* and *end*. Among the different optional parameters, the *header* must specify the type of document, indicating whether the Web browser can display it or a helper program (extension of the browser or a separate program) has to be run. For example, if the type of document is *plain text*, the browser visualises it in an HTML page, but if it is a *model/vrml* document a VR browser is activated. The *body* of the script may contain computations, connections to other servers, instructions to create a document on the fly or various combinations of these. The *end* of the script is a signal for the Web server to close the server-client connection.

Since the Web emerged, it was enrolled to share 2D GIS data. At the current stage, there are two types of Web 2D GIS: server-side and client-side. The server side is based on connections to GIS servers by CGI scripts. In practice, the user requesting GIS information, activates a CGI script, which transforms the query to a GIS executable operation, processes the result in a form appropriate for display in a Web browser and delivers it to the client-site. The result is usually displayed as text or image in an HTML page. Examples of such access to GIS are VISA International, MapQuest. The major disadvantage is that interactivity on the client site is on a very low level: a new pan or zoom operation activates a new connection to the server. The apparent advantage is that all the computations are performed on the server side, which permits employment of existing GIS and DBMS servers.

The client-side GIS is realised following different approaches: *plug-ins (helpers)* and *Java applets*. The plug-ins are programs that enable the Web browser to read and visualise types of data not supported by HTML, i.e. they have to be installed permanently on the client site. Java applets are programs written in the Java language that reside on the server. They are delivered to the client site on user request and stay active for the time of their running. Since the processing of data is shifted to the client site, the possibility for the interaction and manipulation of data is greater. In general, Java applets are considered the more powerful approach that plug-ins, since the applet preserves most of the Web functionality, has a smaller size then an HTML page (the Java applet is a compiled program) and provides flexible means of visualising graphics (e.g. Java 3D). The often-mentioned disadvantage is the restriction of possible connection to one server, i.e. the server where the applet is launched. An example of a GIS plug-in is MapGuide and a Java based GIS is the ArcView Internet Map.

VRML: The Web Standard for 3D Visualisation

The history of the VRML started in 1994 with the first attempts of leading vendors to establish a Web standard for 3D graphics. The language passed through several stages, i.e. VRML 1.0, VRML 2.0 VRML97, before it was endorsed for an Internet standard in December '97. The first edition, VRML 1.0, was capable of describing static 3D models and linking them with VRML, HTML documents. The second version added interaction, dynamics, scripting and a multimedia interface. The key design criteria, i.e. *authorability, composability, extensibility, be capable of implementation, performance, scalability* (Web3D Consortium Inc.1999) aim to:

- enable development of computer programs to create, edit and maintain VRML (i.e. VR modellers)
- enable development of computer programs to visualise and navigate through VRML (i.e. VR browsers)
- encourage development of programs for the conversion of other graphics file formats into VRML (i.e. converters)
- provide means to use, combine and re-use dynamic 3D objects within VRML
- provide the ability to define new object types not explicitly defined in VRML
- enable implementation on a wide range of systems
- ensure interactive performance on a variety of computing platforms.

Despite its short history, VRML has become very popular among software developers. Plenty of VR software (browsers and modellers), as *stand alone* (work as individual application program) or *plug-ins* to Web Browsers, are already available on the market or on the Web. However, the developers of VR browsers still cannot react to the fast changes and improvements of VRML. The functionality of only few browsers (e.g. *CosmoPlayer*, *WorldView*) is according to the specifications.

Designed for a standard Web protocol, VRML enables hyperlinks to other types of documents supported by the Web, e.g. movie, sound,

HTML, VRML, and programs (CGI scripts, ECMAscripts and Java applets). It is capable of representing geometry, texture, materials, lighting, i.e. all the components of the scene, as well as, the behaviour of objects and dynamics. The dynamics ranges from techniques to play complex animation to means of detecting user actions and consequent events. The ability to represent real worlds and their dynamics places VRML among the second-level modelling languages such as Open Inventor. Since the level of realism and dynamics that can be achieved in VRML is very high, the 3D models are called *worlds* or *virtual worlds*. Each world can be almost an unlimited composition of smaller worlds distributed on different servers all over the Web. All these features make VRML an attractive alternative for a front-end interface to 3D GIS data on the Web.

The 3D scene designed according to the VRML syntax is actually an ASCII file format, which can be read within any word processing software. In this text, the 3D scenes described in VRML will be referred to as *worlds* or *documents* (a general term for any data transferable on the Web). The term *VRML model* is frequently used to indicate individual objects (or groups of objects) that can be embedded in a world, e.g. a VRML model of a telephone. Visualisation software, i.e. a VR browser, displays the 3D model. We must distinguish between what can be described inside VRML files and what VR browsers are expected to supply. The VRML provides the parameters for scene description and the animation of objects while, the VR browser cares of rendering the scene and supplying the means to navigate through and interact with the model. Initially, the basic function of the VR browser, besides visualisation, was only real-time navigation, i.e. provision of virtual reality techniques: *examine, fly-over, walk-through, pan, zoom*. The second edition of VRML granted the VR browser new responsibilities related to various dynamics inside the scene.

The VR browser parses the VRML instructions in an internal model (*scene graph*) and ensures the rendering, the detection of user actions and the enabling of the hyperlinks. In the part of tasks related to the visualisation, the VR browser follows the traditional visualisation schema. That is to say the modelling of VRML worlds does not require knowledge of rendering algorithms, windowing, device tracking or the design of tools for interaction. The separation between pure visualisation problems and the 3D modelling of worlds is considered by some authors a critical point that may save programming time and efforts, and ensure better concentration on specific tasks. The authors recommend VRML for systems, where the emphasis is more on the construction and analysis of the application models than on visualisation.

Despite the rapid development of many different VR browsers, only a few of them have satisfactory characteristics. To date, most of the browsers are designed for Windows 95/NT. Macintosh, Unix, OS2 platforms are just occasionally supported. Still many browsers support only VRML1.0, others fail in the implementation of VRML 2.0 functionality. For example, parts of the standard syntax (e.g. *proto, collision, extrusions, inline*) are not supported by many browsers. An obvious bottleneck is the support of scripts. The VRML specifications does not impose a particular scripting language, i.e. it may be one of the Java, ACMAscripts or VRML scripts, which might not be supported by the browser. The way in which the scene described in VRML is rendered varies as well. It seems many vendors optimise their browsers for speed rather than for rendering quality. The awkward difference in path specification (some browsers accept "\" others "/" in specification of *URL's*) of files very often causes failure to access documents. The VRML specification does not treat the question of user interface. The developers of VR browsers are free in the design of tools for interaction. As a result, the user experiences a variety of provided interfaces and functionalities: from the obligatory VR means to *fly-over*, *walk-through*, *examine*, *pan*, *go-to panels*, to various animations, corrections of navigation speed, rendering modes, lights, etc.

Summary

The chapter provides a basis for designing a 3D GIS. As any information system, 3D GIS needs a data model to structure the information and a user interface to access and query the data. The GIS data model comprises data and rules about real objects and more specifically about spatial objects. The GIS as a system for the maintenance of spatial and non-spatial objects is still a wish. Thus the information per object, which is currently stored in the GIS model is related to the geometric and semantic properties of objects, their spatial relationships and time (in spatio-temporal GIS). The geometric characteristics of an object tend to define the object with respect to location in space, shape and size. These characteristics are closely related to abstractions about space and objects. The vector representation allows a more accurate description than the raster representation and therefore it is used for this research. The real objects are commonly associated with four types of objects, i.e. points, lines, surfaces and bodies. Spatial relationships are needed to perform a large variety of GIS functions. The spatial relationships can be represented by several methods; among them, topology is widely applied due to ensured neighbourhood information and invariance under topological transformations. The semantic characteristics of objects and the time are beyond the scope of this thesis and therefore are not discussed here.

Three stages of GIS data model design are distinguished, i.e. conceptual, logical and physical. The GIS can be developed from scratch and then the three design stages have to be completed. The GIS can also be built on an existing database management system, making use of already provided operations. The second approach will be adopted here. Therefore, only two stages are of particular interest for this thesis, i.e. conceptual and logical design. Conceptual design refers to the clarification of objects, their attributes and relations. In this respect, the object-oriented approach and its common principles facilitate the abstraction of real objects. The logical design then is related to mapping the selected components to the relational data model. This process can be favoured by the utilisation of conceptual models (ERM, IFO) that provide methods of describing objects and their relations, and of mapping onto the relational data model.

The user interface is a very important part of 3D GIS development. In general, 3D GIS implies an expensive task in each of its aspects, i.e. data collection, design and maintenance. Therefore the user interface has to permit comprehensive utilisation of the information. The advances in computer graphics offer new extended tools for the visualisation, interaction and manipulation of data, which have to be fully employed for 3D GIS. Queries and visualisation of spatial analysis have to be graphics-oriented rather then text-oriented whenever appropriate.

An important factor in a 3D GIS is remote access to the data. Recent developments on the Web, i.e. VRML make possible visualisation and interaction with 3D models. The exploration of the 3D worlds in the VR browsers ensures a certain level of virtual reality techniques. Existing VRML, HTML and other Web standards and software modules allow the development of GUI with limited effort, relying on some operations provided already by browsers. Therefore VRML and HTML will be employed as a front-end engine to the 3D GIS model.

Investigation of User Requirements for 3D GIS

A question often facing the development of an information systems concerns the origin of the approach followed, i.e. technology- or application-driven. The issue is rather mixed. On one hand, users are the most aware of tasks they deal with, and it seems logical to rely on their requirements. On the other hand, being tied by legal, policy and long-time traditions, and confronted by organisational, financial and market problems, users can hardly identify revolutionary new requirements. The technology-driven approach is not the optimal solution either: the system developed may appear to have extra (or insufficient) functionality for a particular application. However, the hardware and software industry has always been the generator of ideas. In historical terms, the powerful hardware and software were born first and after that 3D models of the real world were created. Apparently, a balance between the two approaches is needed: user problems and routine operations have to be studied to be able to offer contemporary, advanced solutions.

This chapter, in accordance with the first objective of the research, explores the readiness and requirements of users of a municipal 3D GIS. Investigating user requirements for 3D GIS is a broad issue, which definitely cannot be clarified in a single study. The study is expected to provide valuable information about the required types of objects, necessary spatial relationships and the spatial resolution acceptable for most of the users. The demand for 3D visualisation is investigated to an extent relevant to the thesis, i.e. components of the scene, which will effect the GIS model. In this respect, the study focuses on needs for realistic (shaded or textured) 3D models.

The gathering and analysis of user requirements is a sophisticated process demanding careful consideration of many factors. A simple questionnaire given to the personnel responsible for a certain information system could create a very subjective view. In contrast, by ignoring conversations with the staff, an important observation gained from long-term experience could be overlooked. To facilitate investigation, a particular strategy is followed. Methods to determine user requirements are studied to select the most appropriate way of identifying items of interest and sorting out their characteristics. Then existing urban projects resulted in 3D CAD or VRML models are reviewed in order to gain information about objects of interest and resolution. The information kept presently in a municipality is studied to outline existing and potential users, and their interests in terms of type of information and outcomes. Last, a supplementary investigation on currently maintained data, spatial analysis and preferences for 3D visualisation, among producers is conducted. An analysis of the CAD projects, the municipal information and the results of the interview motivate the selection of basic objects of interest, resolution, spatial relationships and GUI for visualisation and interaction that are relevant for the thesis.

Methods of Studying User Requirements

Plenty of methods regarding requirement determination are discussed in the literature. Most of them are business-oriented and related to overall analysis of processes in the organisation (company, firm, agency), starting from the mission and ending with the final outcome. All the five components of a GIS, i.e. hardware, software, people, data and procedures are considered and estimated. In this respect, the scope of this study is limited to only data and procedures (operations). Among them all, two frameworks are relevant for our goals, *Performance, Information, Economy, Control, Efficiency, Services* and *Object-oriented requirements determination.* The first approach refers to the production process, the second approach addresses analysis of items of interest and their interrelations.

Performance, Information, Economy, Control, Efficiency, Services (PIECES): The framework emphasises important questions that become part of a consequent discussion on a requirement's determination. *Performance* concerns the information system needs in order to "work" for the user, e.g. needed response time, printing quality, needs for a GUI. *Information* focuses on the data that the system has to maintain, e.g. input information, output information, the layout of the output, required standard screens. *Economy* addresses the project development, the operational cost and eventual benefits and savings, e.g. appropriate workable solutions, anticipated cost savings, "how the manual activities will be affected by an automated solution". C*ontrol* is associated with the security and editing of data, e.g. at what level (user, screen, database, field, etc.) the security and editing are demanded. *Efficiency* is measured at three levels, i.e. organisational, departmental and individual, e.g. "what value is added to the process", "how the operations will improve the work in the office". *Services* focus on issues related to functional requirements, system maintenance and staff training, e.g. required operations, required training courses. The framework embraces six important aspects of an actual production process and it is highly recommended to gather requirements to upgrade existing information systems.

The PIECES framework does not offer a strategy for resolving data organisation (objects, attributes, relationships). The actual work within the framework starts with a stage where knowledge about similar processes and solutions is required. Moreover, the method is highly related to real effects (in terms of money) if a given solution is implemented. Although improvements in 3D data maintenance and low-cost solutions are intended, benefit/cost effectiveness is not a major objective of the research.

The principles of the framework, however, are critical for the organisation of data in the GIS model and the system architecture. For example, the response time of data traversal to create a VRML document (performance) is closely related to the type and structuring of geometric objects. The remote deleting of an object (control) may be restricted by a specific behaviour hosted in the database. Hence, the framework is to be used for gathering supplementary information on *how* the data has to be organised.

Object-oriented frameworks follow quite different approaches, i.e. the items of interest are the prime focus. Essential advantages of object-oriented frameworks, which contribute to the goal of the research, are the high emphasis on the information, the ability to concentrate on separate components of the system development (e.g. data organisation, data management, data interaction) and the ability to identify objects and clarify their characteristics in a very broad sense.

Coad's object-oriented requirements determination: Coad's object-oriented framework concentrates on three issues, i.e. *objects*, *responsibilities* and *scenarios*. The scope of objects is in practice unlimited: an object can be a physical item, process, action, etc. *Responsibilities* focus on broad sense characteristics by asking three basic questions.

The first question, i.e. *what does the object knows about itself*, tries to collect "personal" properties of the object. The second question, i.e. *who does the object know* pursues the clarification of relationships. The third question, i.e. *what does the object do* comprises objects functions or behaviours. *Scenario* refers to the time-ordered sequence of object interactions or changes in object. In the GIS context, the three questions of responsibilities can be interpreted as questions to investigate attributes, relationships and behaviour. Scenario then can be associated with the time component of GIS. Hence, the framework covers the scope of the GIS content, i.e. it clarifies *what* data has to be in the database.

Although the framework is a successful guide in the information exploration, it pays little attention to implementation issues as they are discussed in PIECES. Therefore, the two frameworks should be combined to explore *what* is to be included in the GIS model and *how* to organise it.

The first two objectives (i.e. objects and responsibilities) of Coad's framework provide a technique to identify real objects, regardless of the nature of the object. The performance (e.g. preferable GUI, demanded realism, data exploration and editing), information (e.g. output information, mostly updated elements) and control (e.g. data protection) are to be employed for visualisation and interaction issues.

The following step is a decision on methods to collect user requirements that reflects the validity and objectivity of the results. Clearly, different groups of people, or documents, or other evidence used, create different pictures and may outline contradictory requirements.

Therefore, the level of subjectivity accepted has to be clarified a priori. Norman 1996 classified the methods of collecting user requirements, with respect to the viewpoint (perspective) they create, into three major groups:

Global perspective: The methods evaluate mostly the experience with existing systems, as the intention is to eliminate the human factor. Such methods are based on: 1) reviews of current reports, 2) conducting of survey on activities already completed by the company, 3) visiting similar system installations. The advantage of these methods is the ability to become familiar with current, established requirements. The main drawback is the concentration on already solved problems, which might be insufficient for future developments.

Individual perspective: The methods aim to reflect and challenge the experience of the user. Interviews, observations, questionnaires, prototypes can be created for the purpose. Although highly subjective, the methods provide useful information on user interfaces, operations and outcomes, which can barely be collected otherwise.

***Group perspective*:** The methods of this group, *group brainstorming, electronic joint application development*, etc., utilise advanced decision-making techniques. Besides some specific differences, the common characteristic is the group discussion on the problem (i.e. user requirements). Advantages can be seen in easily identifying the conflicts among users, the ability to observe reactions to some solutions, etc. The main disadvantage is related to selecting the right people for the group.

In this research, the methods of exploring requirements belong mostly to the first two groups, i.e. global and individual perspective. Since the studied area is rather new, experience regarding similar solutions can hardly be investigated. However, a study on existing 2D and 3D information maintained in municipalities and related institutions delineates the scope of the data currently available. Thus the methods from the first group (revision of existing systems) can still be of help. Supplementary interviews and questionnaires yet within companies that deal with 3D spatial data may reveal essential needs not recognised on municipality level.

In summary, the exploration of user requirements is organised under Coad's object-oriented and PIECES frameworks on the basis of global and individual methods for gathering requirements. The consortium report on a project on a 2D GIS implementation plan (Croswell et al 1994) is used to create a global viewpoint about the current status in a municipality in Bulgaria. A questionnaire on 3D was prepared and distributed among 15 companies (in Bulgaria) with the intention of evaluating individual angles.

The companies form a representative sample of producers selected with respect to 1) the dimensionality of the maintained data (at least 2,5D) and 2) the application orientation (urban or mixed). In addition, similar investigations concerning 3D models supplied valuable information about world wide tendencies. The investigation results will be discussed in the following sections.

Real Objects of Interest in Urban Areas

Since the chapter discusses user requirements, the groups of *users* will first be clarified. Here, a user means a *GIS user*. A particular GIS system can be utilised to collect, maintain, analyse and produce data. However, some of the system users may be responsible only for data collection, others for data storage and analysis. To distinguish between all the users, we will introduce the following terminology: *information centre*, *producer*, *consumer* and *end user*.

Table : GIS users

GIS users	*Data Acquisition*	*Maintenance*	*Query and analysis*	*Production*	*Usage*
Information centre	-	y	y	y	-
Producer	y	y	y	y	-
Consumer	-	y	y	y	y
End users	-	-	y	y	y

An information centre is an institution responsible for the storage, query and analysis of data, as well as for the provision of the result in an appropriate format. The functions of the producer are to collect and process data, and maintain and provide information to other users. Apart from storage, maintenance and production, the consumer benefits from the data in solving some problems.

The end users are not involved in the pre-processing and maintenance stage, i.e. they use the system to query data and utilise the results. If all four categories of users work with the same information system (i.e. GIS) to handle their activities, they are GIS users. Thus, the consumer and the producer are small information centres as well. Note that here the producers are not determined with respect to their selling capacity but with respect to their data collection activities. A municipality is a typical example of a GIS consumer. Other categories that do not have direct access to the GIS will be referred to as non-GIS users. In this context, the requirements of GIS users will be explored.

Inventory of Real Objects

3D real objects of interest for urban applications are often specified in the literature as the scope varying with respect to the application. The common understanding is that the most important real objects in urban areas are buildings (Grun and Dan 1997, Kofler and Gruber 1997, Tempfli 1998b). The CAD 3D models of urban areas (called *city models*) created so far basically consider buildings and terrain represented as TIN.

Fuch 1996 presents a study on real objects of interest for 3D city models, based on questionnaires sent to 55 participants from Europe. The interest in five groups of real objects is investigated: buildings, vegetation, traffic network, public utilities and telecommunications. The results have clearly shown the prevalent usage of (need for) buildings, traffic network and vegetation. Razinger et al 1995 present a virtual model of a square in Graz (created upon municipal request), containing buildings, traffic network (streets and tram railways) lamp-posts and trees. Dahany 1997 suggests three groups of objects to be considered: terrain, vegetation and built form.

In general, most of the authors address real objects with spatial extent. Operational data needed for urban planning in 2D, however, sometimes goes far beyond the real objects of interest discussed so far. Investigation of the organisational and information structure of a municipality (in Bulgaria) has revealed the following specifics: 1) plenty of spatial and non-spatial items stored under different descriptions (DBMS, GIS and CAD) and 2) complex interrelations among different types of data and institutions. For example, personal data are stored in several information centres: three municipal offices and the regional tax office. Is a person an object of interest? How to specify relationships with other objects?

The information needed to reconstruct a number of buildings is hosted in four institutions: the municipality departments, the Electricity company, the Telecommunication company and the Water and Sewerage Corporation. Each of the companies has its own structuring and coding of data, e.g. *deed* is an object according to the Department of State and Municipal Properties, while it is an attribute in the Municipal Land Commission. Is the *deed* an object of interest or an attribute of the parcel? The Department of Architecture maintains plans for urban development, which are extremely important for any new planned construction activity. How should the plans or information inside be referred to: as objects or as the future status of existing objects? A detailed description of the records can be found in Croswell et al 1994.

The following sections present a grouping of real objects that allows the items of municipal interest discussed above to be sorted.

Geometric characteristics of real objects are the leading criterion of the grouping. There are objects with either 1) a complete geometric description, i.e. position, shape and/or size, or 2) only position, or 3) without geometric characteristics. In this respect, we introduce four basic groups of objects to be considered for municipal administration: *juridical* objects (e.g. people, institutions, companies), *topographic* objects (e.g. buildings, streets, utilities), *fictional* objects (e.g. boundaries) and *abstract* objects (e.g. incomes, taxes, deeds).

- The first group of real objects has semantic characteristics, e.g. name, age, status, occupation. The geometric characteristics shape and size are not required; however, the location (e.g. permanent address) is essential.
- The second group comprises all the real objects with discernible boundaries. We assume here that objects with originally indiscernible boundaries have become discernible after pre-processing. The members of the group have complete geometric (i.e. size, shape, position) and semantic characteristics.
- The third group has the same properties, i.e. semantic and geometric characteristics, but their spatial existence is fictive. Examples are neighbourhoods or regions with special status (a centre, industrial areas, residential areas, etc.) or areas with different populations or districts with various levels of pollution.
- The fourth group is abstract objects such as deeds and documents, which do not have geometric characteristics.

All the objects have semantic characteristics presented by their attributes and non-spatial interrelations with other objects. For example, a building has an owner(s) and a corresponding legal document(s), i.e. the real object building interrelates with objects from the juridical and abstract groups. All departments and related institutions keep track of changes in geometric or semantic characteristics. The real objects essential for current 2D GISs are topographic and fictional objects. The real objects from the abstract and juridical groups are maintained in various DBMS or digital and paper files. The table is organised according to the layers as they are maintained in GIS. Although, many of the topographic objects elected from the questionnaire are predefined by state instructions, the results are similar to those already discussed for a municipality.

Table : Real objects in 2D GIS

Layers in 2D GIS	***Real objects***
General map reference	Grid, streets
Administrative units	Districts and suburbs
Parcels	Parcels and sub-parcels, public transportation, rights–of-way, public vending areas
Regulation plan	Project development boundaries
Buildings	Buildings, floors

Road-related paved areas	Paved street areas, parking places, pedestrian walkways, sidewalks
Street centrelines	Centrelines of street segments, delimited by intersection points
Project development plan	Project development outlines
Water distribution system	Water pipe segments, valves, hydrants, service tabs, service lines
Sanitary sewer system	Sewer pipe segments, manholes, valves, service taps, lateral lines
Electric distribution network	Primary and secondary above and under ground electric lines, poles, transformers, switches, fuses, substations, streetlights
Gas distribution system	Gas pipe segments, valves, service taps, service lines, cathodic protection device
Storm drainage system	Aboveground drainage channels, underground drainage lines, culvert openings, catch basins
Telephone network	Centrelines of telephone conductors, switching centres, poles, service lines, other point objects
Topography	Contour lines, height points

The two groups with spatial extent, i.e. *topographic* and *fictional* delineate *spatial* objects. Whilst the topographic objects are primarily considered in the scope of 3D GIS, fictional objects are still not in a focus. The fictional objects are created to serve either global urban development (e.g. regulation plan, development plan) or environmental analysis (e.g. districts for monitoring pollution) or administration management (e.g. suburbs). Maintenance of fictional objects as objects in the 3D GIS-model might be relevant for monitoring urban processes and activities related to such objects, e.g. monitoring noise or air pollution.

The two groups that do not have geometric description (i.e. their spatial extend is not of interest), i.e. juridical and abstract, are to be regarded as *non-spatial* objects. The importance of non-spatial objects for a municipality is apparent. A variety of tasks might be based on the analysis of both spatial and non-spatial information, for example, the distribution of children of primary school ages to be able to plan the locations of new school constructions. In the currently used organisation of information, the names of the children and their addresses will be first obtained from a DBMS and then they will be localised in a GIS system on the basis of their address. In this respect, the maintenance of spatial and non-spatial objects in one information system will result in a significant facilitation.

The issue has another aspect. The non-spatial object does not have spatial extent, however, its status may change in a near future. Recent developments in computer graphics and virtual reality supply techniques to represent a human body (*avatar*) that moves in virtual worlds, speaks and interacts with other virtual persons. For example, an urban planner may participate in a virtual dispute represented by a personal avatar, which can be designed to bring significant messages, reveal characters and preferences objects. Most of the chat-rooms on the Web already offer a library with images for personal identification. Similarly, abstract objects might be represented by media metaphor, e.g. virtual sheets stored in a drawer, etc. One can imagine that the design of geometric and physical properties (colour, material of paper) of historical documents will soon be attempted. That is to say that real objects represented and maintained as non-spatial objects might be transformed into spatial ones in the near future.

In this context, we advocate an integrated GIS capable of maintaining spatial and non-spatial objects. Moreover, 3D GIS should be able to maintain non-spatial objects with their relations, as well as interrelations with spatial objects. This thesis deals in detail with spatial objects, whereas non-spatial objects are considered at a conceptual level or, in exceptional cases, for the sake of completeness.

Geometric Resolution

Here, *resolution* refers to the smallest detail that has to be represented geometrically. Although familiar from 2D, the issue is more sophisticated in 3D and reflects the complexity of geometry. The third dimension poses questions about elements that are insignificant for 2D representations. For example, in 2D GIS the footprints (or roofs) represent buildings and no attention is paid to windows, doors, facade ornaments, etc. As a 3D representation, a building can be associated with a simple rectangular box, or with a composite of several boxes indicating windows, doors, etc. Either all the ornaments on a facade might be considered or only outlines of the wall. A tree might be constructed by several solid primitives, indicating thumb, leaves and branches or by a 3D symbol. Moreover, the criteria for resolution might be different for the different elements of the real object. For example, the ground floors of buildings in central areas (mostly shops) may be require higher resolution than the upper floors (mostly private apartments).

Most of the 3D city models constructed have very simple (low-resolution) geometric descriptions. The usual objects of 3D city models, i.e. buildings, are represented by their walls and roofs. More rarely windows, doors, small balconies, levelled streets and pavements are reconstructed. Quite a lot of models are mapped with photo images. Detailed models including stairs, columns, rooms, furniture, etc., are designed only for individual buildings, e.g. Congress Centre, Graz, Music Centre, Enschede. Fuchs 1996 reports high interest in roofs, as the size of the smallest element is critical for representing front details and overhanging elements. Floors are among the components where demand is very low. Internal constructive parts, e.g. rooms, corridors, are not investigated in the study of Fuchs. The resolution of the 3D city models is an indication of the resolution demanded in 3D. It should not be forgotten that 3D city models fulfil the requirements for exploration, training, guiding, and planning tasks, but not for analysis.

Table : Real objects of interest for 3D

Real objects	***Resolution***
Buildings	Basic: roof, wall, floors Optional: doors, windows, facade ornaments (e.g. building with historical importance)
Bridges	Outlines, columns
Underground	Outlines
Streets	Road-bed, pedestrian areas, parking lots
Parcels	Outlines
Parks	Patches inside the park, significant trees
Vegetation	Trees
Water & sewer network	Segments, connections (inside, outside the buildings), outlines of rivers and lakes
Electricity network	Segments, connections (inside outside, the buildings), traffic lights, lamp-posts
Telecommunications	Segments, connections (inside outside, the buildings), transmitters
Other man-made objects	Monuments, man-made holes
Districts	Boundaries of districts, grids
Terrain	-

Our questionnaire aimed at more detailed exploration of the issue. The results, however, were a bit unexpected: most of the answers in the questionnaire refer to instructions and user requirements. The investigation among the producers exposed several factors that have impact on the resolution maintained: 1) the application (the wish of the user), 2) the chosen method for geometric description, 3) the complexity of the data acquisition procedure, 4) software and hardware for maintenance and 5) accuracy of the source data. The application is the crucial factor for the resolution, e.g. a mobile telephone company may be satisfied with a box as an abstraction of a building, a utility company would prefer to think of a building as a composite object with boxes for each room. The method for geometry description (vector, raster, CSG) influences the detail when one has to stick rigidly to a predefined method for some reason or another (e.g. most of the data are already in this format). In this case, it could be impossible to present some details and the resolution of the model will be reduced. In many cases, the application demands a higher resolution that existing methods cannot ensure or can ensure but only at a very high cost. In this case, the user is restricted due to lack of effective technologies for data acquisition. Software and hardware availability appeared to be an essential consideration for the resolution maintained. For example, the producers of information can easily obtain complete information about the building elements (doors, windows, corridors, rooms, etc.) from construction plans but maintenance of such information is still quite a difficult process due to the large amounts of data and low speed of visualisation and interaction.

Spatial Relationships

From our experience, a systematised study on demand for 3D spatial relationships is not yet available. Therefore quite a lot of effort was spent clarifying the subject. The strategy followed with the questionnaire aimed to: 1) study the software used by the firms, in order to gain information about the type of relationships currently maintained, 2) detect the most frequent 2D analysis applied in daily work and 3) investigate spatial analysis that might be important for 3D. The software used by the firms is mostly 2D GIS with 3D extensions for visualisation.

As can be seen, priorities on metric, semantic and mixed analysis are given. Here, mixed analysis means query of spatial data composing a semantic condition, e.g. "show all the administrative building". The majority of the firms consider buffering analysis quite important as well. The results very much reflect the experience gained from the 2D analysis presently carried out in the offices, e.g. a frequent operation is the buffering of a railway.

Exploring the need for 3D spatial analysis appeared the most difficult task. Personal discussions revealed that firms found neighbourhood and network analysis important for 3D as well; however, they are still not aware of the benefits of 3D solutions. One argument is based on the preferences of non-GIS users (e.g. citizens) for paper maps with results rather than digital copies or screen displays. Another argument is the still quite high demand for digital 2D maps (about 50%). An analogue observation is reported in Fuchs 1996: 10% of producers and 10% of users operate only on digital 2D maps. We can summarise our observations concerning 3D analysis as follows:

The user has a strong tendency to think in 2 or 2.5D concerning spatial analysis. For example, a query "how many meters of pipes are necessary from the street to the 5th floor?" is modified to 1) "how many meters of pipes are necessary from the street to the footprint of the building?" and 2) "how many meters of pipes are necessary for five floors each 3.50 m high?".

- The user does not have any examples of a functional 3D GIS. In many cases, one can hardly picture spatial operations performed in 3D.
- The user is highly influenced by the level of functionality offered by the software in use.

In conclusion, the exploration of user experience in spatial analysis has supplied information about the set of operations that has to be preserved in 3D. Considering the results of the overall discussion above, the objects and the resolution of interest, a meaningful set of 3D spatial relationships can be delineated for the research. The objects of electricity, water and sewage, and telecommunication networks are combined in a group *utility* for simplicity, as some specific objects are explicitly mentioned (e.g. transmitters, lamp-posts). Verbal notations represent the spatial relationships, as discussed with the user.

- Buildings – building *adjacent* (common wall, edge, roof facet) to building; building *adjacent* (common edge, common point) to pavement, street, park, parcel, parking lot; floor, wall *part of* building; window, door *part of* wall; window, door, floor, wall *inside* building; building *inside* building (e.g. garage *inside* house); building *around* park, building

- Bridges – bridge *adjacent* (common surfaces) to park, pavement, street, path; bridge *over* street, pavement, park, parcel; bridge *over* building, bridge
- Streets – street *adjacent* (common boundaries, points) to pavement, park, parcel; street *under* bridge
- Underground – underground *under* pavement, street, park; underground *under* building, underground
- Parcels – parcel *adjacent* (common boundaries, points) parcel, pavement, street, park; building *inside* parcel; tree *inside* parcel
- Parks – park *adjacent* (common boundaries, points) to pavement, street, parcel, path; tree, lamp-post *inside* park, building *inside* park
- Utilities – utility *adjacent* (common point) to utility; utility *adjacent* to wall, terrain, floor; utility *inside* building; lamp-post *part of* electricity network, utility *over* street parcel; utility *under* street, parcel; connection *part of* utility; connection *part of* wall, floor; connection *inside* building; connection *under* street, pavement; transmitter *on* building
- Others – man-hole, monument *inside* pavement, parking lots, parks; tree *inside* man-hole
- Vegetation – tree *inside* parcel, pavement, street, parking lot, parks
- Districts – building, parcel, street, park, utility, monuments *inside* district
- Terrain – surface analysis.

As can be realised, interrelations among objects and their details are to be considered. The argument is that the user may need to distinguish, for example, between the number of windows on the north and on the south facades. In terms of relationships, this requires *inclusion* between walls and windows to be provided.

Realism

The term *realism* is intuitively associated with techniques for visualisation, however, it is related to the resolution and the representation of physical properties (roof material, colour of facades, street surfaces, etc.) of real objects. Thus, the required realism can be achieved either by very high resolution and/or a higher similarity between real and modelled physical properties. An increase in resolution to improve the realism is apparently the more expensive and hence less acceptable solution. The research gives preference to resolution determinated on the basis of application needs and a realism obtained on the basis of a realistic representation of physical properties. Thus the issue becomes significant for the thesis with respect to parameters necessary to ensure the demanded level of realism. For example, if the GIS users prefer shaded models, parameters to indicate the colour of surfaces are sufficient. Relying on technology availability, the users were stimulated to think in terms of utilisation of images to "wrap" the surface of the objects. Many authors have argued that urban models can benefit from photo texturing in several directions: 1) representations of details without geometric modelling, 2) improvements in the orientation while interacting with the model, 3) facilitation of user perception of sizes and shapes. Therefore, the users were asked to consider real photos and artificial images to wrap and map the geometry. Our study on texture needs exhibited high percentages in support of photo texturing. Even those who prefer 2D visualisation and/or 3D wire frame visualisation found only photo texturing meaningful.

Apart from the pursued realism, most of the methods for visualisation that are mentioned are considered quite suitable to display objects, emphasise important object characteristics, control and edit geometric representations:

- Wire frame graphics is still the preferred manner of data editing. Moreover, combinations with algorithms for hidden line (face) removal can be appropriate for controlling the reconstruction of the model, e.g. for consistency checks like "sinking" and "flying" objects.
- Shading methods are better known to the producers working with CAD and multimedia software, which is basically not designed for real-time interaction. Anyhow, the "solid" perception of shading methods is appreciated for interactive manipulation.

Discussions with firm representatives created the impression that the basic CG methods can be applied in a variety of combinations depending on personal preferences, size of data visualised, type of output information, etc. For the research, we will consider the realism an "altering variable". Therefore all the CG methods, i.e. shading, texture mapping and draping will be taken into account.

GUI, Interaction and Operations

The last item in our investigations attempted to determine some aspects related to GUI: mostly manipulated spatial data and the necessity for remote access to data. The first aspect provides information about the objects that have to be accessible for manipulation and thus it influences the geometric representation of data. The second aspect is related to the system architecture for assessing and visualising data. Analysis of the customers of the municipality, as well as the results of the questionnaires, clarified a critical aspect of the issue. An organisation dealing with data acquisition and the frequent update of data, i.e. GIS producers demand extended means to operate on smallest constructive element (e.g. point, line and polygon) of an object. All the participants in the interview belong to the group of producers. VR modellers are not mentioned by any of the firms, which could be an explanation for the higher interest in wire frame. Moreover, producers prefer a stand-alone or Intranet realisation of 3D GIS without a direct connection to other companies. The motivation refers to specific problems relating to the higher speed of interaction with the models and the manner of supplying outputs (i.e. end-product) to customers.

The municipality information system serves internal and external users and in this context the municipality is an *information centre*. On another hand, the municipality governing body is an operand of the information system and hence the municipality is a consumer. However, the municipality certainly does not act as a *producer* of data. Consequently, vast editing and updating operations such as among producers are rarely observed. Instead, visualisation and modifications of individual objects or parts of objects can be expected quite often. For example, discussions on the renovation of a neighbourhood may result in a slight shift in streets or buildings. At this stage, the construction plans are either already approved or not yet available. Hence, the level of editing provided by CAD products (i.e. access to points and polygons) is not required. The process of municipality governing is less interactive than the process of factual design of buildings, streets, etc. Although the urban planners participate in the design, their tasks have a more global scope than the tasks of architectures. For example, de Vries et al 1997 classify the design techniques applied in architectural modelling into: *sketching*, *drawing*, *simulation* and *managing*. *Sketching* is related to conceptual design when no explicit shape is required, the *drawing* stage refines the geometry, *simulation* introduces behaviour (thermal, structural, human) and *managing* refers to the manipulation of data flows (scheduling of the building process). Access to basic constructive primitives is required only for drawing, while for sketching even the primitives can vary. Whilst sketching and drawing are techniques outside global planning interests, simulation and managing may involve environmental impact considerations. These design techniques, however, need means to operate on objects rather than on constructive elements.

The municipality usually has contacts with a variety of users from different organisations, which are equipped with different hardware platforms and software. The common way of exchanging data is digital or paper copies (e-mailed or post-mailed) of needed information. To shorten this process, i.e. save time, effort and money, the municipality has to be able to offer a basic set of operations and data to all organisations needing their service. For example, the members of a telephone company have to be able to check on-line the owners of buildings or ask for statistics about suburbs that are not available in the company in any time during the discussion on a new project. Many of the municipality consumers are regular citizens asking information. Although, hard-copy outcomes will still be popular for quite long time, the importance of on-line services increases every day. The day is coming when computer corners for self-service will replace counters in the municipality. Hence, the municipality has to be capable of supplying information to remote and local users. In this respect, utilisation of the Web is an alternative that ensures flexibility (variety of data formats), accessibility (no location restrictions) and hardware independence (wide availability of Web browsers) of the GIS. Accordingly, the municipality system has to be prepared for a wide range of users with diverse backgrounds. This imposes also GUI and security requirements. The interface has to be user-friendly, flexible enough to cover a large spectrum of questions (in both semantic and geometric domains), and to offer sufficient tools for understanding the results and exploration of the model. The system must have reliable security protection against crackers or unintentional mistakes.

The thesis concentrates on the issues discussed above with some simplifications and limitations:

Operations: Query and display of information are prevalent operations for many activities, as the displayed data can be either text or graphics. For the thesis, this fact implies that means for pointing, formulating queries and visualising of text and 3D graphics has to be ensured.

Interaction: A broad view of the levels of interaction, as the advantages of VR techniques for exploration were discussed. Among the different techniques mentioned, the entry level will be tested for a municipality 3D GIS for a number of reasons:

- Special hardware equipment is not necessary (i.e. mouse and keyboard are the input devices) and hence everyone can benefit from virtual reality techniques.
- WWW standards and protocols to access remote data support the entry level.
- Navigation and exploration supported by VRML and VR browsers do not require specialised skills, and hence, they are quite appropriate for a wide audience of consumers.

Cost: The final words address financial aspects. As mentioned above, the entire process of 3D data acquisition is still quite laborious and expensive. Consequently, the cost of using the information is quite high. If the users (especially the remote ones) are forced, in addition, to purchase high cost hardware and software, the success of the municipal information system is doubtful. Many municipalities all over the world rely on budgets, which as a rule are insufficient. Therefore, among all other considerations, the eventual implementation cost of a municipal 3D GIS should not be overlooked.

User Requirements for a Municipal 3D GIS

The study of a municipality organisation, questionnaires among firms producing spatial information and the worldwide experience in building 3D city models have made it possible to clarify important aspects of user demands relevant for the scope of this thesis. A municipal 3D GIS has to be able to:

- maintain spatial and non-spatial real objects
- provide semantic, geometric and physical characteristics of spatial objects
- maintain 3D spatial relationships
- support interrelationships among spatial non-spatial objects

- formulate spatial and non-spatial queries
- provide the means to perform minor editing
- permit remote access
- offer GUI appropriate for wide audience.

Summary

The chapter presents a commentary on user requirements for a municipal 3D GIS as the scope of objects, their geometric details and spatial relationships important for the thesis are delineated. The subject is very difficult to investigate due to a number of reasons: the high production cost of 3D models and therefore little experience; the lack of functional 3D GIS, a lot of information kept in 2D GISs (spatial analysis traditionally completed in 2D), large amounts of data and complex visualisation (needs of conceptually new software).

The questionnaires among several production companies with urban specialisation provided valuable information on the importance of real objects, demand for spatial analysis and required resolution. Questions related to preference for 3D visualisation helped in clarifying some aspects of realism of urban models and needed operations. A study on the processes and tasks of a municipality specified the scope, distribution and responsibilities for information, basic functions, current and potential partners and customers, present format of outcomes, etc.

Considering the items of interest, a subset of the most important objects (which will be considered in the thesis) is outlined. Two important aspects related to municipal data and tasks were observed: 1) often interrelations between spatial and non-spatial objects are needed, and 2) specific spatial objects without physical appearance in the real world are currently maintained. To distinguish between the objects, a classification regarding the spatial extent, which distributes the objects into four groups, is introduced. Besides the group of topographic objects, a second group with spatial extent, i.e. fictional, is formed. Thus the content of 3D GIS for a municipality is extended to comprise spatial and non-spatial objects. The scope of spatial objects is extended to comprise fictional objects, taking into account their importance for municipality governance.

The investigations gained knowledge of required GUI and outputs. The users agreed on the significance of shading and photo texturing methods for 3D modelling. Thus, the organisation of image files for texturing becomes an essential issue for the GIS model. Texture draping, shading and wire frame graphics are to be used as supplementary techniques to speed up visualisation and improve understanding of the 3D scene.

Several essential conclusions were made on the basis of a specific characteristic of a municipality, i.e. diverse customers having different experiences with GIS data and located in local and remote offices. The visualisation and exploration of outcomes, powered by virtual reality techniques (fly-over, examine, walk-through), will facilitate access to the objects in the scene, speed up exploration of the model and improve readability of the results. The utilisation of Web techniques to access data may fast become a solution for a remote, standardised access.

Visualisation and Interaction Strategy

This chapter covers issues specified by the second objective of this thesis, i.e. it is devoted to system architecture for query and visualisation on the Web. The idea for the approach is inspired by recent developments on the Web. This chapter employs the Web mechanisms to provide remote access to 3D GIS.

The chapter is organised into three parts: 1) a discussion on VRML, 2) an introduction to the system architecture and 3) problems for dynamic creation of VRML worlds. VRML is a carrier of the scene parameters needed for the visualisation and interaction and therefore the chapter provides a detailed review. VRML syntax is discussed to specify whether some of the parameters (and which) have to be stored in the database. The software for creating VRML worlds, and their suitability of urban data is discussed in order to select the most appropriate way for automatic building of VRML worlds. The visualisation software (i.e. VR browsers) is investigated to detect to what degree the principles of VRML specifications are implemented and what are the problems. The issue is related to the correct visualisation.

An approach to retrieve and update 3D spatial and non-spatial data over the Web is introduced. The basic operations (i.e. select object, query and manipulate) needed by the users are then discussed in detail to exhibit the feasibility of the suggested approach. An extended commentary on essential factors related to the dynamic creation of complex VRML worlds (i.e. ensuring means for composition of queries and display of spatial results) aims to clarify whether some parameters require permanent storage in the GIS model. Specific visualisation requirements derived from the VRML syntax, visualisation software and the client/server architecture are formulated at the end.

VRML Concepts

The VRML syntax is based on structuring units called *nodes*, which consist of: 1) the node-type (i.e. *name*), 2) parameters (*fields*), which characterise the functions of the node, 3) special fields for input and output events (*eventIns, eventOuts*), which control dynamics in the scene and 4) the *body* of the node, which contains particular values or other nodes. More than 50 standard nodes provide means to define scene and dynamics.

Several nodes are devoted to the design of the scene, i.e. description of geometry (regular and irregular shapes), illumination (directional, spot, point and ambient lights), materials and textures (draping and mapping of JPEG, GIF, PNG image file formats). The node dealing with geometry, colours and textures, i.e. the *shape node,* has three basic sections: *appearance*, *geometry* and *dynamics*. The*appearance* node cares about the "solid" perception of the shapes, i.e. material (diffuse and specular colours) and texture. Since the concept of simple illumination and shading models is adopted, most of the browsers provide only Gouraund and Phong shading. "Attaching" texture to surface of the objects is supported in two ways: 1) simply wrapping a surface with an image, and 2) texture mapping. The texture mapping requires a second *node* to be specified, which contains the image co-ordinates, corresponding to the co-ordinates of the geometry. The *geometry* of 3D objects can be described by using predefined primitives (cone, box, sphere, etc.) or by sets of *faces*, which are represented by *co-ordinates* of points. The node maintaining *faces*, which is important for the thesis, is presented by two sections: 1) co-ordinate section, which contains all the point's co-ordinates composing an object and 2) description section, which holds an ordered lists of points constituting *faces* that border objects. The normal vector of each *face* is computed on the basis of the *right-hand rule*. For example, the normal vector of a wall with anti-clockwise orientation of the points has a direction toward outside the building.

Combinations of other *nodes*, i.e. *sensors*, *routes* and *interpolators*, introduce dynamics. *Sensors* detect either viewer actions (e.g. mouse move, click, drag), or time changes, or viewer position (visibility, proximity, collision). *Routes* direct the captured event to *interpolators* to alter some *fields* (colour, position, orientation and scale). This mechanism provides mostly direct animation, which is often insufficient to describe complex actions. *Script* or *proto* nodes refer to internal or external scripts and user-defined nodes. In the case of complicated movements and manipulations. *Script* nodes, i.e. ability to execute Java and ECMAscript programs, supply the user with a tool to design his/her own sensors and interpolators, and thus expand functionality of the VRML. An extension toward access and connection to other applications and servers can be achieved by using CGI scripts (will be elaborated further). All the VRML *nodes* can be aggregated in various complex hierarchical composites and altered together. More details about VRML syntax can be found in Ames 1996 and Web3D Consortium 1999.

Visualisation of Urban Models in VRML

Plenty of VRML worlds are already designed and hosted on the Web, e.g. to facilitate shopping, to provide tourist and historic information, to ease the understanding of scientific models (geological, biological). Tempfli and Pilouk 1996 report the first utilisation of VRML as a "walk-through" engine of an experimental 3D GIS. Kofler 1996 presents a successful walk-through large 3D VRML worlds. Ogao 1977 reports a study on the suitability of the language for cartographic visualisation. Gahegan 1998 employs VRML techniques to enhance exploratory visualisation of geological objects (transparency, focus attention, etc). Schickler 1997 reports a photo-textured model of Atlanta Airport with two different LOD to simulate the taxiing system. Bodum 1998 utilises VRML for visualisation of municipality information. Doyle et al 1998, concentrate on an interesting aspect of VRML implementations, i.e. virtual presence in the design process by *avatars*. The variety of experiments with VRML worlds is an indication of the increasing interest in the language for end visualisation.

However, the potentiality of VRML and VR browsers to manage dynamics and interact with the model is still underestimated. VRML and VR browsers are generally understood to be a system for the visualisation of 3D graphics on the Web allowing real-time navigation. The specific manner for providing dynamics and maintaining interaction contributes to this opinion. It is important to realise that the dynamics introduced has to be described in advance. If one wants to be able to click on a building, a sensor has to be attached to this building in the VRML document. If one wants to have the animation of a walk on a street, the route and the speed of walking have to be specified in a special VRML node. The VR browser provides, in addition, the freedom to move inside the world while the particular dynamics runs. These features of the VRML concept are further explored for an experimental GUI for 3D GIS. In the following text, we will elaborate on the positive and negative characteristics of VRML, with respect to the visualisation process of a 3D GIS of urban areas.

Means to describe urban worlds. Between the two geometric representations (i.e. face and primitives), preference has to be given to face description. The predefined primitives provide a simpler and more compact than face description. However, GIS models maintaining topology are based on boundary representations, i.e. they operate with faces, points, arcs, etc. Moreover, most of the 3D models are reconstructed from surface measurements that are difficult to combine with simple shapes (cone, sphere, cone). Another reason refers to the texturing of the geometry. VRML supports the two mechanisms to attach image to geometry, i.e. texture mapping and texture draping. The more precise adjustment, i.e. texture mapping, is permitted only for surfaces (composed of faces) or individual faces. The syntax is such that permits the usage of only one image file per VRML node, i.e. textures for a separate surface or an individual face have to be available in one image file. If, for example, one face has textures of different images for its two sides, then the face has to be described in two VRML nodes. However, there is no way to map, for example, six different images onto the six sides of a predefined shape cube. The operation in this case is texture draping, i.e. the image covers the entire shape according to a rule. Thus, an accurate texturing of the images requires face description of real objects. Texture draping may be quite an appropriate technique for large surfaces represented by a set of faces, e.g. terrain. The geo-referencing between each polygon (i.e. triangle from TIN) and the corresponding piece of image (required for texture mapping) is not efficient: 1) requiring algorithms to generate image co-ordinates and 2) almost doubling the size of the VRML world.

The use of faces to describe urban objects inevitably leads to a need to group/separate objects, which can easily be achieved by *Group* and *Transform* nodes. Since the roof and the walls are represented by faces, they can be textured with separate image files. Often, real objects used to be represented as lines and points in the GIS model.

Although VRML support descriptions of lines and points, utilisation of predefined primitives is recommended. For example, lines and points can be displayed as tiny cylinders and little spheres. The substitution usually increases the readability of the VRML world. The VRML standard lines and points frequently cause "disappearing" of the object while navigating through the model. The effect is observed when the line object lays exactly on a face.

Means to describe large worlds : The *Inline* node offers a solution to the organisation of large data, typical of urban areas and often denoted as an obstacle to visualisation. The idea is that a very large VRML world can be parsed into several small worlds, which can be hosted separately (even on different servers). The *Inline node* maintains the reference to the worlds and their relative position to each other that allows the VR browser to assemble them into one. The node is quite popular for introducing predefined VRML models in virtual worlds. For example, models of a desk, a telephone, a fax machine, which are organised as individual VRML documents, can be located at different places in a room. Some models with quite complex geometry, sensors and scripts are already available on the Web for free utilisation and distribution. From the GIS aspect, the node may be successfully used for symbolising real objects such as trees, cars, planes, etc.

The *LOD node* permits LOD techniques to be applied during visualisation in a simple and flexible manner. The syntax of the node requires the different representations and the switching parameter. The different geometric descriptions can be organised per object (or group of object) and organised in one or several VRML worlds. In practice, each object can have several representations in separate VRML worlds. Since each LOD is a regular VRML document, the construction of the LOD becomes entirely the designer's wish. LOD might be created either for geometry or texture or both. For example, frequently used LOD for buildings are: *very detailed*, e.g. all geometric details of the buildings plus texture; *less detailed*, e.g. only geometry without texture; and *coarse*, e.g. only outlines of the buildings. VRML specifications agreed one parameter, i.e. the distance between the object and the viewer, to switch between LOD. The VRML worlds created so far usually have predefined LOD organised in separate VRML documents (Schickler 1997).

Despite the appropriate means to create 3D urban scenes, the language has a number of missing features and drawbacks:

- Topology per object: As mentioned before, the face description is based on 2-manifold subdivision of space, which ensures topology per object but not among objects. Thus, the representation cannot be used directly for spatial analysis, other than ones derived on a metric (co-ordinate) basis. For example, the query "are the two buildings adjacent?" (which can be organised in a Java script) will provoke a check for common co-ordinates. In this context, the VRML can be used only as an intermediate step between the 3D GIS model and internal structuring of the browser.
- Lack of a database interface: Originally, VRML was not thought to allow access to databases and thus an SQL node is not included in the syntax. On the basis of recent investigations and experiments, an elaborated syntax of the SQL node inside the current version of VRML was suggested.
- Lack of semantic descriptions of models. VRML specification does not intend maintenance of semantic information, which consequently requires the employment of other (e.g. HTML) standards.
- Lack of means to manipulate objects: Although providing *sensors, interpolators* and *routes* to simulate interaction (will be discussed), the truth is that the language lacks editing tools. The browsers are not expected to save the VRML world either.
- Large documents: The VRML world is uncompressed ASCII text, which in practice results in very large documents. Length has an impact on the time for delivery and parsing, i.e. the display of the file on the screen. Some of the VR browsers support a *gzip* compression, which reduces the delivery time a bit.

The description of the scene is the first part of the visualisation process. Yet the VRML document has to be displayed on the screen by the VR browser. The experience gained with VR browsers lets us identify various limitations. The first problem is related to the visualisation of faces with more than three vertices. Although VRML allows an arbitrary number of vertices per face, the rendering engines used by browsers (mostly OpenGL, Direct3X) require triangles. Non-triangulated faces influence the display in two directions: wrong visualisation of the face and incorrect display of the spatial relationships of two objects (e.g. a line that is supposed to rest completely on a surface might fly over it). This means that the browser has to activate algorithms for the triangulation of non-triangles. Two types of triangulation are provided: for concave and convex (default) faces. The choice is made on the basis of a flag in the VRML world. Since the algorithms for concave triangulation are time-consuming and their permanent usage is not recommendable, there are two alternatives: 1) execution of the triangulation algorithms prior to the design of each VRML world, and 2) database storage of concave faces.

Another problem arises in the visualisation of holes. Based on the properties of 2 manifolds, the rendering engines need "opening" of the holes, i.e. connecting to the bordering face. Not all the VR browsers can handle holes, but the capable ones require opening. The operation can be achieved by a special ordering of the vertices. The special order combined with a flag for concave face will enable a correct visualisation. A similar problem arises in the visualisation of overlapping or hosted faces. Similarly, to the visualisation of concave faces, an appropriate partitioning of the faces can ensure the correct visualisation.

Approaches to Design VRML Worlds

Software to generate a VRML world in its complete functionality is still to come. Since the birth of VRML, many vendors have already extended their file format to create VRML; others have started development of the software, based only on the VRML world format. To date, several types of computer programs can be used to create VRML worlds. While the first group of software only exports, the second group exports and imports VRML worlds. The VRML worlds created by CAD and GIS are only scenes for visualisation, lacking the dynamics and possibilities for interaction.

The second group comprises various software, which can be divided into three groups, i.e. *VR modellers, VRML editors* and *VRML modellers*. *VR modellers* (*Medit, Pioneer Pro*) provide extended tools for the design of illumination, shading models and texture mapping, viewing points, and camera positions, and they create richer VRML worlds than the first group. Most of them provide plenty of texture and shape libraries with means to map textures, modify and deform objects, create animation, etc., which allows fast and efficient design of 3D models. Unfortunately, the VRML nodes introducing dynamics (*interpolators,sensors, routes)* are not usually supported. *VRML editors* are extended text editors to manually type the body of the VRML world. Similarly to some HTML editors of the same range, they facilitate the user by furnishing the syntax of the standard VRML nodes. Although they support all the *nodes,* the method is not efficient because of the rather long time to create a single VRML world.

Perhaps, the most convenient way of designing VRML worlds is provided by *VRML modellers*, e.g. *V-Realm, VR creator, Cosmo Worlds.* This software operates directly on the VRML nodes, i.e. the user is able to modify the fields of the nodes. The GUI supplies windows for both graphics and text representations of the VRML world, as the individual solutions vary.

As can be realised, software to create VRML worlds exists in many variations. The extended GUI is the apparent benefit of their utilisation for display, interaction and modification of VRML worlds. Unfortunately, they are hardly appropriate for our goals. First, most of the modellers are stand-alone solutions and none of them is designed for access over the Web. Second, they do not have a connection to a database, i.e. the product of the modelling is a graphics file. Third, the maintained topology (in the modellers and VRML) is based on a 2-manifold subdivision that requires a 3D topology creation after each change in the VRML world. An eventual utilisation of such software would require the development of a number of extensions: 1) translators from and to the GIS model, and 2) remote access to databases. Moreover, each user accessing the system must have installed the modeller and the extensions on his/her computer, which will effect the overall price of the system.

System Architecture for a Web 3D GIS

The number of client/server architectures utilising VRML is still very limited. Some of the drawbacks of VRML and VR browsers listed above, the intensive improvement of the VRML specifications (three versions in four years) and the prevalent usage of VR techniques for entertainment (e.g. computer games) are some of the factors influencing the client/server realisations based on VRML.

Lindenbeck and Ulmer 1998 present a server application of VRML for geological surfaces and objects. The surfaces represented as triangular meshes are stored in related groups of files on the server. The VRML document is created on the fly, according to a prior selection of data (i.e. triangular meshes). The selection is formulated in an HTML form and processed by a CGI script. In fact, the options provided in the HTML form correspond to the names of the files that contain the meshes. The GCI script reads the geometry from the selected files and creates the VRML document. In this respect, the query completed is a query on server files rather then a spatial query.

The role of the VRML document stops at simple visualisation i.e. complex geological structures are displayed in different combinations to facilitate the visual investigation of surface intersections.

The systems presented by Dodge et al 1998 and GIS technologies 1999 make connections to 2D GIS (ESRI) and visualise some information in dynamically created VRML documents. The utilisation of VRML is primarily for 3D visualisation. Since the third dimension is created by extrusion of a parameter (geometric or thematic) of the 2D database, the 3D models lack of vertical details. More elaborated use of VRML is intended in the research of the IGG, Rostock 1998. VRML documents are employed to query thematic attributes (e.g. eaves and ridge height, usage of floors) of objects.

Coors and Jung 1998 report the results of a prototype system (GOOVI-3D) for the interactive query of spatial and semantic information, and the visualisation of the results. The approach, based on CORBA-ID and Java applets, establishes direct access to databases inside the VRML document. The information is stored in two databases for geometry and topology, and semantics as the ID of an object is the link between them. The server accesses at least one of the databases to complete the result of query. The communication between the server and the client is controlled by Java applets, i.e. the function of the VR browser is only the execution of the applet. In this respect, the system architecture is client-oriented. The SQL query, which is specified in a specially designed SQL node (defined in a *proto node*), is conducted by Java applets. The result of the query is a set of ID of objects. The ID set is compared with the ID of existing objects (named nodes) in the VRML document and those that are equal are modified, e.g. highlighted. For this purpose, a second node is constructed to associate a node from the VRML document with the ID of an object. To manage the link between object and VRML node, an extension of the transfer protocol is developed. The VRML document is delivered initially as a script, which controls the relation between ID of objects and VRML nodes. Thus the user can identify the objects, query them and visualise results. This approach seems very promising but still cannot be widely implemented because it requires the specialised Web protocol IIOP.

Query, Visualisation, Modification

The system architecture used in our approach is a server-oriented, i.e. the CGI mechanism is utilised to access remote information. Among the approaches to access data on the Web, preference was given to the CGI scripting due to the well-experienced cross-platform mechanism, standard HTTP protocol, and availability of freeware API to access databases and create HTML fill-out forms. The proposed system architecture is intended for evaluating and verifying data organisation on the server and therefore less interest is paid to the functionality on the client site. The overall structure of the system comprises a Web browser with a VR plug-in on the client site and a Web server and a database system on the server site. The system relies on HTML documents for the composition of queries and visualisation of data other than 3D graphics. Therefore a plug-in instead of a stand-alone VR browser is required. VRML documents are intended for identifying (selecting) objects to query and visualisation of spatial analysis. CGI scripts establish the protocol between the client and the server. They are responsible for the assembling of SQL queries, the access to the RDBMS and the creation of documents (HTML or VRML) on the fly with respect to the result of the query. Although in principle similar to the approach presented by Lindenbeck and Ulmer 1998, our system intends more sophisticated tasks for CGI scripts, HTML and VRML documents and thus goes a step further toward the analysis (spatial and semantic) of data.

The CGI mechanism with its specific client/server protocol subdivides the process of query, visualisation and modification of data into several different stages:

User identification and database selection : In the first stage, the user is given options from which to chose a model (e.g. each town or even neighbourhood might be in a separate database) and the range of interaction (query and visualisation, updating) has to be clarified. The database can be selected inside an initial HTML document (by filling in a form, pointing to an area of interest on a 2D raster map) or VRML document (pointing on a 3D map). The most convenient way is to provide each user with a number of default operations to query and visualise. Thus each user can send queries and receive responses (HTML or VRML documents) without restriction, except a change of data in the database.

Query: The request for information per object (a particular building, an owner or a list of co-ordinates of a parcel) can be organised in two steps: first, the object has to be identified, and second, the type of the information has to be specified. For example, the simple query "who is the owner of this building?" will require: 1) the means to point to the building and 2) an appropriate interface to specify the needed information. The first step can be formulated either in a VRML or an HTML document. In the HTML document, the query can be specified in a fill-out form (multiple choice or directly typing SQL queries). This means that the user has to be aware of the ID of the object in advance, which is often impossible. The alternative, i.e. pointing the object in the VRML document, is much more attractive. Despite the lack of a real pointing interface, combinations of sensors and scripts simulate the same operation (to be discussed in Section 4.2.2). The user visually chooses the object of interest as in a standard CAD system or GIS.

Both ways of identifying an object activate a CGI script on the server. The script delivers an initial HTML form, where the user makes further clarification and sends it back to the server. First, the form parameters are processed, the data needed according to the request are extracted, a document is created and then sent back to the client station. Since the CGI client/server architecture is stateless, each new query initiates the same process and results in a completely new document created on the fly. Very often, however, some information from already filled-in forms has to be forwarded to the next query. In this case, the information (some values of particular parameters) has to be memorised. Since the server has no memory of previous connections, parameters are kept on the client station. The intermediate document created contains the needed parameters in "hidden" for the user fields and passes them back to the server with the new query.

The two-step schema described above is appropriate only to query information (non-spatial and spatial) about a particular object. Many queries and much analysis cannot fit in the schema due to impossibility of clarifying objects in advance. Examples of such queries are "show the highest building in the town", "show all the administrative buildings", "show the common walls", " who are the owners of the buildings along this street?". The composition of such queries is rather too complex to be organised in VRML documents. Therefore special HTML fill-out forms, where the user types either the necessary SQL statement or other appropriate parameters, have to be created and hosted on the server. A detailed discussion is given in the next paragraph.

Data visualisation : The information delivered at the client site is displayed either in an HTML document (text, 2D graphics, etc.) or in a VRML document (3D graphics and text). For example, the query "show the way from the hotel to the nearest shop for shoes " will result in a subset of objects (streets and surrounding houses) that can be displayed in a VRML document. Animation may even route the user from the hotel to the shop. The result of a more complex request "show the way from the hotel to the nearest shop for shoes and the prices of the shoes" will be separated into two steps. First, the VRML document with the geometry will be displayed. A second user action, e.g. click with the mouse on the shop, will take care of the creation of an HTML document delivering information about prices. This limitation is related to the CGI mechanism to create documents on the fly. The first line sent by the server is the MIME type of document. The creation of two documents on the fly is impossible.

Local query and manipulation means temporary changes in the data on the client station. The local query can be understood as the exploration of some properties that are described in the VRML or HTML document, but can be activated and/or visualised only by user action. For example, one may want to compare several architectural plans for the reconstruction of existing facades. The existing facade will be stored in the database and the user will operate with several new views of the facade available as a separate image files on the server. Another typical example is the design of vegetation and the evaluation of its future development. Different types and sizes of trees can be prepared and send to the client and the user can compare and evaluate the effects in different views. The way to make such temporary designs of geometric information is provided by the VRML, i.e. the suite of sensors, interpolators, and embedded scripts. The interpolators and scripts run only on the subset of data at this particular moment at the client station and, consequently, do not influence the information on the server, nor other clients.

The resulting documents can be displayed on the client screen as one window with several frames, several new windows, or combinations of them. In the prototype, system preference is given to one Web window, split into two or three frames. Separate windows provide the user with more freedom to resize and adjust observed models; however, the control on displayed information is disturbed, which leads to chaos on the client desktop.

Bibliography

Arrhenius, Gustaf: *Evolution of the Solar System*, Washington D.C. NASA, 1976.

Baerbel K. Lucchitta: *Multispectral Landsat Images of Antarctica*, Washington, 1987.

Baker, J. A.: *Liberating Life: Contemporary Approaches to Ecological Theology*, Maryknoll, NY: Orbis Books, 1990.

Carey, S. Warren: *Theories of the Earth and Universe*, Stanford, Stanford Univ. Press, 1988.

Carleton, Andrew M.: *Satellite Remote Sensing in Climatology*, London, CRC Press, 1991.

Dyer, M. I., and D. A. Crossley: *Coupling of Ecological Studies with Remote Sensing: Potentials at Four Biosphere Reserves in the United States*, Washington, 1986.

Edward A. Beaumont and Norman H. Foster: *Remote sensing*, American Association of Petroleum Geologists, 1992.

Edwards, Denis : *Earth Revealing Earth Healing: Ecology and Christian Theology*, Collegeville: Liturgical Press, 2001.

Friedhoff, R. M. and W. Benzon: *The Second Computer Revolution: Visualization*, New York, 1991.

Ghassem Asrar : *Theory and Applications of Optical Remote Sensing*, New York, Wiley, 1989.

Gorkavyi, N. N.: *Physics of Planetary Rings: Celestial Mechanics of Continuous Media*, Berlin: Springer-Verlag, 1999.

Hall, Dorothy K., and Jaroslav Martinec: *Remote Sensing of Ice and Snow*, London, New York, Chapman and Hall, 1985.

Hall, Edward T.: *Early Stockaded Settlements in Goveraiador*, New York: Columbia University, 1944.

Hall, Edward T.: *Spatial Analysis in Archaeology*, Cambridge: Cambridge University Press, 1976.

Jamwal, Anil: *Principles of Remote Sensing*, Jnanada Prakashan, Delhi, 2010.

Johnson, Cait : *Earth, Water, Fire and Air: Essential Ways of Connecting to Spirit*, Woodstock, VT: Skylight Paths, 2003.

Kennie, T. J. M. and M. C. Matthews : *Remote Sensing in Civil Engineering*, Glasgow, Surrey University Press; New York, 1985.

Kuhn, Thomas S. : *The Structure of Scientific Revolutions,* Chicago: University of Chicago Press, 1970.

Lindgren, David T. : *Land Use Planning and Remote Sensing*, Canada, Kluwer Academic Pub., 1985.

Mack, Pamela Etter : *Viewing the Earth: The Social Construction of the Landsat Satellite System*, Cambridge, MIT Press, 1900.

Measures, Raymond M. : *Laser Remote Sensing: Fundamentals and Applications*, Malabar, Fla., Krieger Pub. 1992.

Oldroyd, David : *Earth Cycles: A Historical Perspective*, Westport, Connicticut: Greenwood Press, 2006.

Philip H. Swain and Shirley M. Davis : *Remote Sensing: The Quantitative Approach,* London, New York, 1978.

Rees, W. G. : *Physical Principles of Remote Sensing*, Cambridge, New York, Cambridge University Press, 1990.

Richards, John A. : *Remote Sensing Digital Image Analysis: An Introduction*, Berlin, New York, 1993.

Schanda, Erwin : *Physical Fundamentals of Remote Sensing,* Berlin, New York, 1986.

Sharieff, Afzal ; Masood Ali Khan and A. Balakishan : *An Introduction to Remote Sensing,* Sarup Book Pub, Delhi, 2010.

Wall, Stephen D., and Kenneth W. Ledbetter : *Design of Mission Operations Systems for Scientific Remote Sensing*, London, Washington, 1991.

Watson, O. Michael : *Symbolic and Expressive use of Space: An Introduction to Proxemic Behavior,* New York, *1972.*

Wolf, P. R. : *Elements of Photogrammetry,* New York, 1983.

Yadav, Surendra Kumar : *Soil Ecology : Remote Sensing and Geographic Information System in Management,* APH, Delhi, 2007.

Index

A

Absorption Spectroscopy, 191, 197.
Aerial Photography, 71, 73.
Agriculture, 179.
Air Ministry, 11, 12, 13, 14, 15, 59.
Analytic Signal, 203, 209, 211.
Analytical Chemistry, 194.
Analytical Techniques, 218.
Archaeology, 176, 222.
Astronomy, 47, 179, 195.
Atmospheric Environment, 177.
Automated Cartography, 108, 109.

B

Biology, 117, 180.

C

Calibration Constants, 188.
Cellular Automaton, 122.
Chemical Types, 131.
Christian Huelsmeyer, 3, 6, 57.
Civil Aviation Radars, 43.

E

Economics, 116.
Elementary Cellular Automata, 125.
Environmental Monitoring, 138.

G

General Description, 173.
Generalized Transforms, 215.
Geodetic, 73.
Geology, 178.
Geometric Resolution, 263.
Geostationary, 133, 135.
Geostatistics, 109, 118.
Guglielmo Marconi, 5, 34.

H

Hydrological Modelling, 107.

I

Imperial Army, 27, 28.
Imperial Navy, 25, 27, 28, 29.
Indian Remote Sensing, 79, 80.
Information Sciences, 116.
Interaction Strategy, 273.

L

Law Enforcement, 180.
Light Meter, 185.

M

Marine Mammals, 146.
Markov Chain, 114, 118.
Mathematical Biology, 117.

Measuring Air Pressure, 172.
Meteorology, 177.
Military, 2, 40, 144, 152, 180.
Military Radars, 40.
Modern Instruments, 162.
Modern Recording, 165.

N

Navigational Technologies, 152.
Networks, 107, 265.

O

Ocean Surveillance, 154.

P

Parametric Sonar, 158.
Passive Sonar, 147.
Physics, 18, 20, 21, 31, 37, 179, 195.
Polarization, 52.

Q

Queueing Theory, 116.

R

Radar Astronomy, 47.
Radiometric Correction, 76, 168.
Radiometric Resolution, 75, 167.
Related Automata, 130.
Related Methods, 112.
Related Terms, 111.
Remote Sensing, 2, 78, 79, 80, 82, 171, 183, 194.
Remote Sensing Software, 78, 79, 171.

S

Scaling Filter, 212.
Seismoelectrical Method, 222.
Seismometer, 158, 162, 164, 165.
Social Sciences, 117.
Soil Science, 178.
Sound Propagation, 141.
Spatial Relationships, 237, 239, 266.
Spectral Resolution, 75, 167.
Submarine Navigation, 152.
Support Vector Machine, 119.

T

Teleseismometers, 163.
Temporal Resolution, 75, 167.
Transponder, 145.

U

Underwater Security, 154.
United States Army, 9, 59.
United States Navy, 7, 56, 154.

V

Vehicle Location, 157.
Visible Light Reflectance Photometry, 190.
Visible Spectrum, 134.
Visualisation, 225, 241, 250, 251, 273, 275, 281.

W

Water Velocity Measurement, 157.
Wave Measurement, 157.
Wavelet, 204, 210, 211, 212, 213, 214, 215.
Wavelet Function, 213.
Wavelet Transforms, 214.
Weather Radar, 44.
Weather Satellite, 133.